D0422572

# The Rise of Climate Science

Kathie and Ed Cox Jr. Books on Conservation Leadership

Sponsored by

MEADOWS CENTER
FOR WATER AND THE ENVIRONMENT
TEXAS STATE UNIVERSITY

Andrew Sansom, General Editor

# THE RISE OF CLIMATE SCIENCE

## OF

## CLIMATE

## SCIENCE

A Memoir

GERALD R. NORTH

TEXAS A&M UNIVERSITY PRESS  •  COLLEGE STATION

This paper meets the requirements
of ANSI/NISO Z39.48-1992 (Permanence of Paper).
Binding materials have been chosen for durability.
Manufactured in the United States of America

∞

Library of Congress Cataloging-in-Publication Data

Names: North, Gerald R., author.
Title: The rise of climate science : a memoir / Gerald R. North.
Other titles: Kathie and Ed Cox Jr. books on conservation leadership.
Description: First edition. | College Station, [Texas] : Texas A&M
University Press, [2020] | Series: Kathie and Ed Cox Jr. books on
conservation leadership | Includes bibliographical references and index.
Identifiers: LCCN 2020007520 | ISBN 9781623498672 (cloth) | ISBN
9781623498689 (ebook)
Subjects: LCSH: North, Gerald R. | Tropical Rainfall Measuring
Mission—History. | Climatologists—United States—Biography. |
College
teachers—United States—Biography. | Climatology—History. | Global
warming—Research—United States. | Climatic
changes—Research—International cooperation. | Global
warming—Political aspects—United States. | Global
warming—Research—Government policy—United States. | LCGFT:
Autobiographies.
Classification: LCC QC858.N67 N67 2020 | DDC 551.6092 [B]—dc23
LC record available at https://lccn.loc.gov/2020007520

A list of titles in this series is available at the end of the book.
Visit  https://atmo.tamu.edu/people/profiles/emeritus-faculty/northgerald.html

# CONTENTS

# FOREWORD

Years ago, when I became executive director of The Texas Parks and Wildlife Department, black bear sightings in our state were a rarity. We might have fifteen or so reports of bears a year, and all of those would be in the Big Bend region. I remember taking a call one day in 1990 from a hysterical woman on a ranch near the border with Mexico who was looking out her kitchen window at a full-grown bear tearing the laundry off her clothesline in her backyard.

Today, we're seeing bears in places like Kerr and Lampasas Counties, far north of the Rio Grande, and the number of sightings has dramatically increased. The movement of birds and animals northward is but one indication of a changing climate but a clear sign that any one of us can observe and understand.

Critical to our ability to address changes taking place in the climate is the tremendous increase in our scientific understanding of the increasingly disturbing phenomenon that has occurred over roughly the same period. In The Rise of Climate Science, Gerald North, distinguished professor emeritus at Texas A&M University, humbly and graciously tells us the story of his own life and his role in the evolving science of climate change.

Rising from simple beginnings, North describes a rich and engaging passage from childhood salesman in a local Tennessee haberdashery to eminent scientist in the field he helped develop through the years. Along the way, a unique blend of his own life history and the science to which he has contributed so much leave little doubt that no story is more important to conservation today than the accurate and defensible explanation of what may well be the greatest threat to the planet in millennia.

What makes North's account most compelling is his tender and candid personal memoir alongside an erudite account of his own journey in the climate space and the advances in the science itself. The importance of this story at this moment in time makes it a fitting and important addition to the Kathie and Ed Cox Series on Conservation Leadership. It is further fitting that Professor North is a distinguished member of the Texas A&M family, which also includes Texas A&M University Press, publisher of the series in partnership with The Meadows Center for Water and the Environment at Texas State University. The Center, with support from The Meadows Foundation, is changing the conversation on climate change with bold new initiatives in research, education, and leadership.

As I write, the sound of white wing doves wafts in the windows from my backyard in Austin. Scarcely thirty years ago, the only place white wings were found in Texas was in the Lower Rio Grande Valley. Now, there are more of them here in Travis County than Cameron County along the border with Mexico.

Gerald North's work in climate science offers an equally understandable illustration of the unprecedented threats and changes facing the coming generation.

—*Andrew Sansom, General Editor*

# The Rise of Climate Science

# INTRODUCTION

This book tells the story of the emergence of climate science as a modern physics-based discipline from its infancy in the early 1970s to its mature status today. The story is told through references from my own journey into and continuing career in the field of climate science: how I grew up in the upper South, my education and teaching as a middling theoretical physicist, how and why I transitioned to climate science from a tenured faculty position in physics, and my maturation as a scientist as I settled into an exciting and rapidly growing field. I do not claim any singularly significant role in the field's emergence, although like many, I have made a number of contributions and certainly made acquaintances and friendships with most of the leading players. I also have had hands-on experience in many of the subdisciplines in climate science, from pencil-and-paper climate models and modern statistical methods to Earth observation systems based in space. Likewise, I have touched base with most of the institutionalized forms of the sciences at home and abroad, from giant government laboratories to large and small universities, in locations ranging from the United States, the United Kingdom, Japan, and the Union of Soviet Socialist Republics. In this book, I relate some stories from those visits with the intention of shedding light on the nature of science at home and abroad during the Cold War period. My employment has ranged from scientific programmer to academic department head and member of boards of trustees of nonprofit organizations supervising climate research.

I tell the story first of my unlikely background and my rather peculiar and spotty array of talents and outsized ambitions. It took me a while to realize that I had a gift for mathematical and statistical models for physics and related problems, while simultaneously discovering the hard way that I was inept in the laboratory. I have attributed the root of some failures to a self-diagnosed but likely mild case of dyslexia that slowed my reading speed and seemed to always make me hit the wrong buttons on keyboards. That handicap made me a terribly inefficient computer programmer as well. I am generally a friendly fellow with a sense of humor, sometimes corny. As a teenager working in retail stores, I learned how to sell products—a skill that has paid off repeatedly throughout my life.

After the historic launch of the Russian satellite Sputnik in 1957, science took off in America as it became evident that a strong scientific presence in our country would be the basis for deterring aggressive moves by our adversaries abroad. Not only physics and chemistry, but derivative fields such as meteorology and oceanography were relevant in this rush to be world leaders in the physical sciences. The federal government began to build institutions that included peacetime atomic power, the space (including missiles) program, reconnaissance satellites, computer-based weather forecasting, and many others. While spinning off practical peacetime applications in these fields, the nation demonstrated its power to a global audience. The National Science Foundation (NSF), the Atomic Energy Commission (AEC, now the Department of Energy, DOE), the National Aeronautics and Space Administration (NASA), the National Oceanic and Atmospheric Administration (NOAA), and the Department of Defense (DOD) all proceeded to create laboratories comparable in size to those of the great public universities across the country. As with military bases, these "national labs" were dispersed across many congressional districts, thereby assuring their own sustainability.

Physical science opened a pathway for young aspirants from all classes and ethnicities, opening and widening the gates for a fresh population of talent to enter the scientific workplace. A new meritocracy was forming as a result of this steady stream of state-sponsored demand. I was swept into this current, body and soul. Tremendous resources in materials, previously untapped people-power, wealth, a can-do culture, freedom of expression, and its unique geography enabled America to quickly grab the lead in the world of science, as evidenced by the number of Nobel Laureates among many other indicators since the end of World War II.

My story also touches the history of science in the Cold War era. I was in the Union of Soviet Socialist Republics (USSR) five times and had a chance to observe that country and the contrasts between their and our ways of science and life. I also tell of my many visits to institutions in Japan and Europe, learning and practicing my craft as I went.

I also participated in and/or witnessed political negotiations in small groups such as university selection committees, government decision processes such as contractor selection committees and satellite mission selections, and larger intergovernmental negotiations about international partnerships.

I will start my story a little out of chronological order by describing a pivotal experience in my life. It is probably indicative of such epiphanies in the lives of many ambitious young scientists of the time. For me, there were many such forks in the road both before and after, but how could I forget this one?

## Boulder and the National Center for Atmospheric Research

October 1974: Looking south down the Front Range of the Rockies from my fifth-floor office I could see Pike's Peak, seventy-five miles due south. Its dark gray silhouette stood out through the background of lighter gray mist. The smog in Denver obscures the optical path connecting Boulder and Colorado Springs. The Peak was not visible on most days, only on especially clear ones, depending on weather patterns. On mornings when I could see it, I would often bring it to the attention of my friendly next-office neighbor, Tom Lundgren, who was interested in air turbulence and a seasoned aeronautical engineering professor on sabbatical leave from the University of Minnesota. I, on the other hand, was evolving from being a middling theoretical physicist to an atmospheric scientist during a one-year sabbatical leave from my tenured position at the University of Missouri-St. Louis (UMSL). The transition was tricky because one develops an intuition as a young meteorologist that is unnatural for middle-aged scientists coming from neighboring disciplines to pick up.

My office was atop the north tower of the National Center for Atmospheric Research (NCAR) in one of the south-facing turrets or "crow's nests" shown in the picture of the NCAR building. The modernist architect I. M. Pei designed this spectacular building, the color and texture of which blends in with the pink sandstone of the Flat Irons, which grace the slopes just west of the city of Boulder, Colorado. Dubbed a "Cathedral for Science," the NCAR Building rests on Table Mesa, six hundred feet above the city, at a total elevation of about six thousand feet above sea level. Just driving every morning up the winding road to the top of what is now called Walter Orr Roberts Mesa was an exhilarating experience. Deer could often be seen along the roadside or grazing just outside ground floor windows of NCAR meeting rooms.

I had been teaching in the St. Louis branch of the University of Missouri for the past six years after a rigorous training to be a physics researcher. While I loved teaching, I was beginning to be uncomfortable with the low upper limits of such a career. At the time, I was hanging on by my fingernails in the overcrowded and supercompetitive atmosphere in my field. For a few years, I had been reading about atmospheric science and even attended a couple of conferences in that discipline. I was too ambitious to do the same job day after day until retirement without really making a difference in my native field—also not sufficiently smart or indefatigable. I had been exceptional in graduate school, so-so as a postdoc, OK as a teacher, but a bit short when cut loose into the adult competition. I yearned for a career in which my contributions would be influential. My wife, Jane, and I had visited Boulder for a summer school of lectures for young scientists in the late sixties, and I thought returning would be fun for the family. We had two children in grammar school, whom I thought would enjoy the experience.

A few weeks after arriving, I came across a manuscript titled "Climate Modeling," a review article being circulated for comments before submission for publication. Stephen H. Schneider and Robert Dickinson, both young *senior scientists* at NCAR, had written the paper. I hardly sensed it then, but these two new friends of mine later came to be recognized as among the founding fathers of climate science, evidenced later by their inductions into the National Academy of Sciences (NAS). Another manuscript came to my attention, written by two other new friends, Petr Chylek and James Coakley, both PhD physicists just ahead of me in their career transformation paths. This paper left open some questions, and I immediately went to work on trying to answer them. I immersed myself into the problem of climate change and in particular simple climate models.

By the end of October, I had settled on a problem based on the papers I mentioned and the many references in them. I benefited from encouragement of permanent senior Advanced Study Program (ASP) members such as Philip D. Thompson.

Another mentor to me was C. E. "Chuck" Leith, who had developed one of the first general circulation models, which he also programmed himself.

Of course, at the time, I was unaware of the seminal contributions these people had made to atmospheric sciences during their careers.

A group assembled every day at around 2:00 p.m. for coffee at a table just outside my office. Phil Thompson was always there. In fact, he often started his day at that time and place; afterward, he worked in his office nearby until late in the evening. John M. "Mike" Wallace, a few years younger than I and a professor of atmospheric sciences at the University of Washington (UW), was located just across the hall. He was on sabbatical, coauthoring a survey-level textbook of meteorology for incoming graduate students at UW. The book was intended for students who had majored in physics or chemistry as undergraduates. He handed me chapters of the manuscript drafts to read and comment on. Maurice Blackmon, another physics professor undergoing the same retread as I, was on sabbatical leave from Syracuse University. Maurice went straight to analysis of output data from computer runs of the NCAR numerical climate model, mainly in collaboration with Wallace.

NCAR gave me complete freedom to work on anything I chose. I took advantage of that opportunity to try my luck on the climate problem. While I did not have the training and intuition of a meteorologist, I had some promising tools in my toolbox. This time at NCAR turned out to be a big break for me. The famous theoretical physicist P. A. M. Dirac once said that in the early days of quantum mechanics, even a second-rate physicist could make a difference, years later as the field developed and plateaued into a later phase, even a first-rate physicist could make only a second rate contribution (Farmelo 2011).

Climate studies became climate science with the creation of climate models. Climate models provided a way to incorporate the laws of physics and applied mathematics in framing the problem in quantitative terms. The problem was to build a simulation model whose solution was to produce an artificial climate that was faithful in representing the present real climate, including its wind and current systems, temperature, and water substance distribution. The model should yield solutions for these variables when some external parameter, such as the Sun's brightness, is changed artificially by the investigator. Several groups around the world began experimenting with these so-called general circulation models (GCMs) using the newly available high-speed computers. Despite promising steps forward in the early 1970s, the available computers were too slow, and their storage capacities were too limited to include enough of the essential details, although they captured the gross features of the wind systems. The grand program of modeling the general circulation of the atmosphere proceeded steadily alongside the progress in computer technology—the high-speed computers got faster, and their storage capacity enlarged with each succeeding year—climate models improved themselves in parallel synchrony, always nearly choking the world's fastest available computers. The growth rates for computer speed and capacity were exponential. The general circulation model at NCAR, under leadership of Warren Washington and Akira Kasahara, was among the world leaders (see Donner et al. 2011).

The keen desire to get quantitative about the past, present, and future climate stimulated the concept of a hierarchy of climate models, discussed in the review paper by Schneider and Dickinson. The bottom rungs of the hierarchy start with the global average surface temperature, which is governed by the balance of the rates of heating by sunlight and the rate of the planet's cooling by sending infrared radiation back into space. This balance can be expressed as an equation, whose solution yields the planet's global average temperature. Simplified "energy balance models," which were more complicated than the global average models but which did not require high-speed computers, emerged in the late 1960s and early 1970s. These "pencil-and-paper" models caught my eye.

I started with climate models based on the pioneering work (first published in English in 1969) of the Soviet scientist Mikhail I. Budyko, then director of the Main Geophysical Observatory (MGO) in Leningrad (now St. Petersburg), and a paper published by William D. Sellers (1969), then professor at the University of Arizona. Over decadal-term averages, energy conservation determines the globally averaged climate at the surface. The Budyko-Sellers models went one step further. They included the latitude as an independent variable. The climate in this simplified scheme consisted of the average temperature around a latitude circle at each latitude. The solution to their energy balance equation was a curve showing the surface temperature versus the latitude—cold at the poles, hot at the equator. Without poleward transport of heat, the model would yield a solution too cold at the poles and too hot at the equator. The energy balance equation had to include a mechanism to cool the tropics and warm the polar regions. Budyko and Sellers chose different mechanisms to achieve this, both finding satisfactory curves of temperature versus latitude. They each had some shortcomings that raised questions—Budyko's model, though elegant, was too simple; Sellers's model had too many adjustable or empirical parameters and so for such a simple idea was too complicated. Both had another curious property: if the Sun were to dim by only a few percent, the planet would make a catastrophic plunge in temperature to become an ice-covered "snowball Earth." This alone was enough to draw serious attention.

I got to work right away on the problem by constructing a model that was somewhere between the Budyko and Sellers models, simple in concept but more physical (linear heat diffusion of heat energy from warm toward cool regions). The energy balance was expressed as a differential equation that ordinarily would have to be solved by computer (as did Sellers's). All three of the models had polar icecaps whose latitudinal extent was adjusted automatically by a separate equation (or constraint), according to the temperature at the ice cap edge. We could turn the proverbial "knob" representing the Sun's brightness to see what happened. If you increased the Sun's brightness, the ice cap would shrink in areal extent and the global average temperature would rise. If you turned the solar brightness down, the opposite would happen, the ice cap would grow. All three models incorporated this feature called the *ice-albedo feedback mechanism*—more snow cover meant more cooling.

My model required more mathematical sophistication to solve. Using the methods in my toolbox, I solved the problem. The analytical solution clarified the differences between the earlier models and opened the path within the same framework to include more effects, such as the seasonal cycle and longitude dependence, perhaps even quantitative, but approximate, answers to puzzles in past and future climates.

As the quotable German theoretician Arnold Sommerfeld said about a century ago, paraphrasing, "If you want to open a lock, you can use a key or a sledgehammer." Sommerfeld was my academic great-grandfather. Four of his own graduate students won the Nobel Prize in Physics, but he never did, although he was nominated many times, perhaps more than anyone. His metaphor of the key and sledgehammer applies in our context to the analytical versus the numerical (computer) solution.

I submitted my paper, titled "Analytical Solution to a Simple Climate Model with Diffusive Heat Transport," to the *Journal of the Atmospheric*

*Sciences* in February 1975. I had found the same "snowball Earth" property discovered by my illustrious predecessors: if the solar brightness were lowered by only a few percent, snowball Earth comes crashing down on us. This paper was mathematically difficult, using what mathematicians call *hypergeometric functions*, leaving many meteorology readers flat—"OK, another mathematician showing off." In a couple of months, I found a simpler and more easily scrutinized way of getting the same results. I submitted that paper, titled "Theory of Energy Balance Climate Models," in May 1975 to the same journal. This one was even more like Sommerfeld's *key* in terms of its mathematical brevity and elegance. The solution expressed in the later form was a rapidly convergent infinite series. The terms are functions of latitude called *modes*, and the sum of only the first two terms was sufficiently accurate to serve as a good approximation to the exact solution (infinite number of modes included in the sum). At about this time, I learned of a paper by two Princeton graduate students, Isaac Held and Max Suarez (1974), that had just appeared in the journal *Tellus*. They found roughly the same result. Yet another paper was soon to be published by NYU mathematics graduate student, Michael Ghil (1976). Ghil's paper used sophisticated numerical methods to study and solve the problem. As in my work, his also considered the stability of the solution roots. We were blazing a busy trail. All three of my young "competitors" went on to sparkling careers in climate science.

My new pals at NCAR conferred all the appropriate handshakes and pats on the back. I still remember the reaction of septuagenarian Bernhardt Haurwitz, a legendary and periodic visitor to ASP who spent a day or two per month at NCAR, away from his professorship at Colorado State University, located about an hour and a half's drive north of Boulder in Fort Collins. He always arranged to join us at coffee break just outside my office. Somehow my ever-present (sometimes lame) humor seemed to rub him the wrong way (I thought), and he seemed cool toward me. Actually, I did not know how very influential he was in the history of modern meteorology. Before publishing my February paper, I gave him a copy of the typed manuscript. After that he was completely different in his behavior toward me. I think he had discovered that I was a serious beginner in what turned out to be a nascent field of science.

My two NCAR papers were published in my thirty-seventh year. This is supposed to be the time that theoretical physicists were finished or at least on a steep downhill slope. Maybe as a theoretical physicist I was finished. However, this fresh start and early taste of success was a very stimulating event for me. I was new in a new field.

## NCAR

Before getting on with my story, I want to say a few words about NCAR because part of this book is about the different institutions that played significant roles in the emergence of climate science as well as in the development of my career. I have already mentioned the rapid growth of science after World War II. After leaving his high-level managerial post at the Los Alamos site where atomic weapons were made, Vannevar Bush returned to his old faculty position at MIT. In 1945 he published an influential report, *Science the Endless Frontier* (still available online). He advised President Truman and founded the NSF. He cited meteorology and oceanography as candidates for growth and advocated for the creation and nurturing of large institutional establishments with facilities that might be shared with university researchers. Many of these institutions were already in place because of the Atomic Energy Commission. I will visit some of these in more detail in later chapters.

NSF founded NCAR in 1960. It is one of several institutions funded mainly by NSF, although now some funding is contracted to NCAR by other agencies. NCAR is operated by a consortium of universities, and it encourages collaboration with academic researchers through its massive facilities, research aircraft, and other resources. NCAR's early mission was primarily to conduct atmospheric research that would improve weather forecasting, in particular, through the use of high-speed computers, mobile radars, and aircraft that would be housed on-site in Boulder. By the time I arrived in Boulder in 1974, the research mission had also expanded to include climate science, mostly through general circulation models that would run on the NCAR supercomputer. NCAR continues to be a world leader in this line of research.

# CHAPTER 1

# APPALACHIAN ORIGINS

## Early Life and Family

I was born in the little town of Sweetwater, Tennessee, on June 28, 1938, with the name Gerald Raymond North. I vaguely recall from my mother that "Raymond" was part of the delivery doctor's name. My father, Paul "Sanford" North, was an auto mechanic at the Chevrolet dealer there. Soon we moved to Statesboro, Georgia, where he was in charge of the repair department at the dealership. At some point, it became clear that America's role in WWII was expanding rapidly, and we were soon to be drawn in. Although over thirty, my father expected to be drafted. In the early 1940s, we moved to Knoxville, Tennessee, where he was in charge of the shop at Reeder Chevrolet. The purpose of the move was to bring us closer to my mother's family, who were situated in and around Knoxville, while my father served on a light cruiser named the *USS Amsterdam* in the Pacific.

In particular, my mother's close older brothers, Jefferson "Glenn" Hill and Clarence Hill, and their wives lived in or near Knoxville. She also had a younger half-sister, Edith, and half-brother, Kenneth Hill, in the area. Kenneth was drafted into the army and served in the Normandy Landing in France in 1945.

On my mother's side, I am descended from Scotch-Irish and English Hillbillies who settled in Appalachia, earlier in Virginia, then western North Carolina, and finally in East Tennessee, the last two in locations inside the present Great Smoky Mountain National Park. My father's people were a mix of Scotch-Irish and German from Dandridge, Tennessee, in the fertile Tennessee Valley, middle-class farmers and later petty bourgeois.

My mother, Marjorie, called me "Jerry," and I called her "Mom." My father called me "Mac," and I called him the same, the nickname used for one another among sailors. I have heard that my parents met when she worked in the ticket booth at the Roxy Movie Theater on Market Square in Knoxville, where the farmers brought their produce for sale off their truck beds. I think some of the movies there were pretty racy. They married in 1936 when my mother was about seventeen. My dad and her brother Glenn were buddies and were consumed with automobiles and motorcycles in their youth. I think they once attended the Indianapolis 500, possibly traveling there on their motorcycles. Both were probably heavy drinkers as young men. Glenn had a terrible car wreck before the war in which some of his scalp was burned along with a part of one ear and a piece of one eyebrow. Even with the scarring, he was a handsome and wonderfully affable man.

Glenn and Jean had no children, and this circumstance might have caused him to be drafted into the US Navy before my father. He served as a gunner on the *USS Raymond*, a destroyer escort (DE-341). In 1944 he saw action at Guadalcanal and the Battle off Samar, and later his ship served as an escort at Iwo Jima and Okinawa.

My grandfather, Green Lee Hill, was friends with Benjamin Joffe, whom we called "Pop," and his Jewish family. Pop owned a dry goods store (the Dixie Store) on Central Street just a block or two east off Gay Street, the main business street in downtown Knoxville. Market Square was a block west of Gay. Pop had immigrated to New York City from his homeland, one of the Baltic countries, before WWI. Violent anti-Semitism had made life unbearable there. Soon after immigrating, he was drafted and sent right back to fight on the Western Front against Germany. After WWI he apprenticed as a tailor in New York and used that skill as a clothier in Knoxville. In his store, he sold, altered, and repaired suits and pants. He also sold all kinds of men's clothing, hats, and shoes, new and used. A man could also pawn a pair of shoes at the Dixie Store on Saturday for a weekend date, redeeming it next payday. Pop was a gentleman, good in business, and had a big heart. I came to love Pop.

As with many "Russian" Jewish immigrants of his generation who started out in New York, they saved their money and migrated with their families down the Appalachians to towns like Johnson City, Knoxville, and Chattanooga. Pop's oldest son, Louis, was born in Kentucky, one of the family's odd stops along the way. Pop had cousins in all the cities just mentioned. There they opened small businesses, and most of them prospered.

Central Street was a district where European immigrant merchants had storefronts easily accessible to farmers who came to town on Saturdays. It was also a favorite shopping area for African Americans who lived in the Knoxville area. A few doors away stood a fresh fruit and vegetable stand owned by Italian Americans and another store by Jewish merchants that included shoe repair. There was a small sandwich shop owned by Greek immigrants. A close friend of the Joffes named Leon Sarof owned a pawnshop on the corner of Central Avenue and Vine Street. In 1959, Leon sold me an engagement and wedding ring package for my marriage to Jane. Because of our family friendship, he made me a good deal.

Diagonally across the intersection from the pawnshop was a pharmacy, which had a platform that opened like a stage on the corner. On Saturdays, a barker-vendor would sell patent medicines there (the vendors barked like country preachers). As a child I loved to stand in the crowd outside that store and listen to the pitch offered by that huckster on the stage. The show often opened with live snakes (somehow related to the peddling of patent medicine) and live country music, then the hard sell began. These ointments and tonics could cure any ailment, from sore muscles to failing male sexual prowess. Actually, my mother did not like me hanging around there and would often yank me away.

Glenn's wife, Aunt Jean, worked at the Dixie Store as a salesclerk all through WWII while Glenn was away in the Pacific. Starting at about age six, I would sometimes visit the store and was drafted into selling men's Rayon socks, draped over an open-topped pasteboard box that was displayed outside the storefront, similar to the fruits and veggies displayed a couple of doors away. Customers thought I was cute with my corn-silk hair. From a very early age, I learned something about business.

## World War II

By early 1944, Uncle Sam had drafted my father into the US Navy and sent him off to Naval Station Great Lakes for basic training. There he learned about electronics and especially the repair and maintenance of radar systems. The electronics course lasted a full year. Radar was new and highly classified. He was thirty-four years old, tall (5'11") and skinny (about 130 lbs.). He never learned to swim but did master the art of floating (especially in saltwater!). He told me that since he was older than the other guys, they let him off easy on the calisthenics during basic training.

My father was very intelligent. He was valedictorian of his high school class of about thirty in Dandridge, and his mother told me she thought he might have had a career as a preacher. He once told me he had read the Bible, cover to cover, twice. I believe it. But he despised the fire and brimstone preachers that were common in the time and place of my youth.

Later as a teenager, I tried to beat him at checkers. I never was able to do it, even when I was in college. I later learned that he was the checkers champion on his light cruiser in the Pacific. Another thing I remember about his intelligence was that he subscribed to both Knoxville newspapers, and he worked both crosswords by the clock, rarely losing to the timer. I tried to do those crosswords but never could get to first base.

Both of the Joffe sons were in the US Army. The older one, Louie, had been a basketball player at Knoxville High School. Louie served in the Aleutian Islands off the Alaskan mainland. My dad said that Louie won lots of money from his fellow soldiers in the Army, possibly a nest egg for his business investment later. His younger brother, Abie, was sent to the European Theater in the Army. After the war, both Louie and Abie opened stores, one next door to Pop's Dixie Store and the other around the corner on Vine. Louie sold new clothing and dry goods, and his brother sold small appliances. Abie later went back to school and eventually became a pharmacist, opening a pharmacy in Oak Ridge, Tennessee. One of the brothers' cohorts was David Richer, whose family were fine fur merchants at a fashionable store on Knoxville's Main Street. When they all came home from the war, there were card games on a blanket spread in the yard outside the Joffe house, and I ran around pestering them.

When Mrs. Joffe died in approximately 1944, my mother, Aunt Jean, and I moved into the house with Pop, who was very distraught with his two wonderful sons overseas and still grieving the loss of his wife. Aunt Jean worked at the Dixie Store, and my mother looked after the rather large house on McCalla Avenue and me. In the front window hung the familiar displays of the four men of the house who were overseas in the war. Happily, none were killed or even injured. There were different colors on the window displays around town identifying in which branch of service someone served, and whether they had been injured or killed.

I still have postcards from my father and Glenn from faraway places. In 1945, my father finished his training and boarded the light cruiser *USS Amsterdam* heading toward Honolulu and then Tokyo Bay. Luckily, just before the ship came close to its destination, the atomic bombs were dropped, and the war was soon over. Nevertheless, the war changed both Glenn and my father. Both accelerated their drinking, and most of their former ambitions were stunted. The naval experience had left a permanent impression on my dad. While it enabled him to see a larger world, he had not sought it. I think being on a ship, going across the Pacific, never knowing when that torpedo might hit, had some traumatic residuals, but he never talked about it.

As an only child, I asked about having a sibling someday; my father always replied, "We cannot afford another child." Aside from our nickname for each other, Mac, he also liked to use naval terms such as the "deck" for the floor and the "head" for the toilet—I used them too in his presence. He always called my mom "Skipper." He referred to Englishmen as "limeys." While he drank too much, he was a sedentary drinker. He never raised his voice to my mother or me, and he never touched either of us in anger. Come to think of it, he never raised his voice to anyone. My mother applied the few spankings I ever received, all under the age of about eight.

I came out of the war period with a loving but overly protective mom—as they say, "I was spoiled rotten." I ate too many sweets, hated vegetables, and was scared of heights, bicycles, skates, and water. I was skinny as a rail and lacked strength—they wondered if I had rheumatic fever, heart murmur, and other weakening ailments. My mother washed clothes in the bathtub with a washboard and ironed the sheets, my underwear, and probably my socks. I compensated in school by being a smart aleck, later a comedian.

After the war, Pop Joffe took me on two purchasing trips for the store, via rail. We slept in our upright coach seats and stayed at YMCA hotels. The first trip was to Chicago, where he bought wholesale goods to sell in his store—shoes, pants, work gloves, overalls, and so on. He took me to the Planetarium, the Field Museum, and the Brookfield Zoo in the western suburbs. We rode the elevated train, and I stood on the windy shore of Lake Michigan. The next year we went to New York City where we did similar things (Bronx Zoo, Empire State Building, and more). We visited some old acquaintances in the city, and he sometimes passed me off (perhaps with a wink) as his grandson—it must have been a hard sell with my light blond hair. These two trips enlarged my horizons, and even at that young age, I felt that I would someday leave East Tennessee for the larger world.

My dad clearly loved me, but he was a very quiet man. He really did not like talking much to anybody. I was always curious about his past, and especially his naval experience, but he was only available for a few minutes of discourse before he returned to reading. He did relate to me one (possibly apocryphal as to his participation) story when he and some of his friends played a terrific Halloween prank. They had a farming neighbor who was not so friendly to kids. They took apart a wagon and reassembled it on the old man's barn roof. What a sight to wake up the morning after Halloween and find your wagon atop the barn! Another prank was that they cut the wires to the school bell with wire cutters that did not break the rubber insolation. This made it nearly impossible to spot the break. It was pretty strange and chaotic when the bell did not ring for class the next morning. These stories are so ironic considering the sedentary lifestyle he led as I grew up.

After the war, my father and Glenn returned physically unharmed, and our family moved into a trailer a block from the Joffe home. My mom rode to work with Pop and worked in Louie's new store as a clerk. My father was the manager of the Western Auto store on Knoxville's main drag, Gay Street. In a few years he left that job, I think because he hated managing people and meeting with customers—the two ways to make money in retail work. He was talented but crippled by anxiety. I do not know to what extent the anxiety was due to his war experience.

He moved to become a small appliance repairman at Sears, and he fixed everything from toasters and radios to (eventually) TVs. He liked being back in the shop with his soldering iron, meters, and small hand tools, only appearing occasionally to brief a customer on the job he had finished for them. As a challenge, he liked to figure out the problem with a radio or phonograph based on the client's description, before performing any tests with his diagnostic equipment. I went into the shop from time to time since we did not live too far away from the big Sears store. He always had a cigarette stuck in his lips or smoldering in the ashtray—which was really the end of a three-inch artillery shell used in antiaircraft fire. It had his name engraved on its side. My dad was good at this job. Sears kept statistics on the efficiency of repairmen at their stores across the South, and he was usually the leader.

## Latchkey Kid

When I was midway through third grade, we relocated to a one-bedroom apartment located above a drug store that sold sodas, magazines, and over-the-counter medicines, but featured no prescription pharmacy. Once an armed bandit came in the store in search of narcotics and had to be convinced that they had none. Often, especially on weekends, teenagers and young adult males gathered on the wide concreted area in front of the store in the evenings. They were just beneath our often-opened window in the bedroom, in which I also slept. There was noise and an occasional fistfight, resulting in a visit from the cops.

I was a latchkey kid. That first winter and spring at our apartment, I rode to and from school on a streetcar down Magnolia Avenue only a few stops from the railway viaduct, my visual indicator to pull the cord for "next stop." The teenage girl in the house on Randolph Street just behind the drug store was employed by my mom to see to it that I was OK after school. My daily chore was to fill a two-gallon container of fuel oil from a fifty-gallon drum behind the building and haul it up the stairs to the apartment. I then had to pour the oil into the stove's small tank and light the pilot flame with a match to warm up the rooms, which were left cool during the day while we were out. Otherwise, I was on my own, browsing in books, "reading" comic books, drawing, creating my own cartoon strips, listening to radio shows such as *Lone Ranger*, *The Green Hornet*, and lots of others. I was not lonely because I had plenty to do. I did not read much—reading and arithmetic were difficult for me—drawing was OK. Eventually, I did find a neighborhood friend, Joe Franklin, who had a back yard, suitable for passing a baseball and football. We played many hours of stickball there, too. The stickball was made of tightly wadded up paper, wrapped with adhesive tape. The bat was an old broomstick handle.

The owners of the store beneath our apartment, the Lawhons, had just bought it when we moved in. They were our landlords, but they also became friends. On some weekend evenings they would bring beer and enjoy sharing it and takeout food with my parents. I would listen from my bedroom to their stories and jokes. Mr. Lawhon was a University of Tennessee, Knoxville (UTK) graduate and manager of the Knoxville Bus Terminal. My current knowledge tells me that he was bipolar and suffered episodes during which he had to be hospitalized. The Lawhons were very kind and fond of me. They allowed me to read comics and magazines— with the condition that I take them from the shelf and go into the back room, so (other) freeloading customers would not see me. The only serious reading I did was "how to do it" magazines on how to play baseball and football, step by step. I learned how to grip, throw, slide, swing, and punt by following the line drawings above the instructions. This paid off in stickball.

## Adolescence

When I got to junior high, we moved to the upper floor of a two-story house on Gratz Street, just a block off Broadway in North Knoxville and a short walk to Sears. In a year or two, we swapped with our landlord, Mrs. Vaughn, a widow who had lived in the downstairs apartment. We had a backyard where I raised chickens (bantam Rhode Island Reds, Polish, Cornish, bantam game chickens, pigeons, and others). I had lots of neighborhood friends who helped me build pens and care for the birds. I also had a few rabbits for a short time. I think we ate them. I even showed some of my chickens in the Knox County Fair, which was held annually in Knoxville at Chilhowee Public Park.

I got the neighborhood kids into building crystal radios—my father provided some assistance: a sketch of the wiring diagram. He told me that he built the first radio he had ever seen from parts bought from a catalog. A crystal radio consisted of a coil wrapped on a heavy pasteboard tube (it took one sturdier than the cardboard cylinder in a toilet paper roll), a condenser (capacitor), a crystal mounted in a lead base (about the size of a penny, only thicker, like a thick pill). A little detector that had a sharp-tipped wire was used to touch the top of the crystal (it acted like a *diode* in today's parlance). Finally, a pair of earphones was hooked up, and an antenna had to be strung up outside the house like a clothesline. This latter was done at the consternation of my mother and the wonder of our landlady. The pieces were mounted with screws on a small (sanded, varnished, and polished) block of wood, the "chassis" of the instrument. The coil was adjustable in length so that you could tune the a.m. frequency to the desired station. Imagine the excitement of the Huck-Finn-like crowd surrounding the receiver when the first station was detected! Each guy, grasping for the headphones. Of course, lots of trial and error was preliminary to the big event.

It was the best time for my dad and me. We lived about a mile from Knoxville's minor league baseball park, and he would walk with me to the games sometimes several times a week during the season. Other times, I would walk with neighborhood kids. We were part of a club called the "Knothole Gang," named after those who watched the game through knotholes in the outfield fence. Club members often got free or discounted tickets. The Knoxville Smokies were a Class B league farm team to the New York Giants at that time. I think they won the pennant that year. A couple of the players (Foster Castleman and Ron Samford) made it to the big league team, but only for a few games in later years. Although not very talented and pretty skinny, I worked hard to play baseball and did play on a team for twelve-year-olds affiliated with the Boy's Club of Knoxville. I was an avid sports fan, and I followed the UTK football team on the radio, but never attended a game. The newspapers showed portraits of the players, and I drew idolatrous sketches of them.

During junior high and high school, I worked Saturdays during the school year and full time in the summers as a clerk at one of Louie Joffe's stores. He eventually owned three stores that catered to the same kind of working-class or farming clientele. I learned to approach a customer, explain our merchandise, help with fittings, and complete the sale. I also lettered the cardboard signs with a wide-tipped pen dipped in India ink. These stood on clip-stands over the counters or tables housing the merchandise. This was a great experience for me since the act of displaying and selling a product is a key skill often called upon in science. The pay was low, perhaps $5.00 for working all day Saturday, but my father, who thought highly of the Joffes, said I should do it even for nothing because the skills learned were so valuable. I think he was right. Many of the customers were African Americans (colored people, as we said then) and I developed a liking for them. Most of the others were working-class whites (overalls, jeans, boots, gloves, etc.), and I was sympathetic to their world. I was a cheerful salesman, just as today. The Lonsdale store where

I worked on Tennessee Avenue stayed open on Friday nights until 9:00 p.m. Louie would join us after he closed the downtown store, and he and I would stay until the Lonsdale store closed. Afterward, he would take me out for a steak at one of the better restaurants in Knoxville.

One minor price I paid for working for the Joffe's was that the neighborhood I lived in had lots of boys with whom I got along, but most were disgustingly anti-Semitic. I was always teased with the nickname "Heeby" or "Abie" from age eleven to fifteen. I tried to ignore it, but it stung. These kids loved me, and vice versa, but teasing was part of growing up. I always tried to explain what wonderful people my Jewish friends and African American customers were, but they never got it. We all used the N-word, but once I learned it was hurtful, I never again used it. Schools were segregated, and my friends had had no personal experience with African American people.

## High School

In about 1952, my parents bought a small, two-bedroom house (1050 sq. ft.) for $6500 at the corner of Columbia Avenue and Harvey Street in North Knoxville just after I started ninth grade at Fulton High School, which had opened only a year before. It was a combination of a trade school and a regular liberal arts school. About half of the kids opted for the trade school curriculum that included auto mechanics, print shop with linotypes, sheet metal, and similar choices. The girls could choose cosmetology, secretarial, cooking, and others. My first homeroom was in the sheet metal shop.

My father met one of my junior high teachers, Mr. Monday, who happened to be a customer at the Sears repair shop. I was in both of Mr. Monday's machine shop and mathematics classes. My father was not only smart but also good with his hands in a shop—I still have a bookcase beautifully wrought of heavy oak that he built in his high school shop. Mr. Monday told him I was pretty weak in the shop, but actually very good in math   by then I had overcome some of my arithmetical shortcomings (which I attributed to a wicked fourth grade teacher). My father thought I should be a medical doctor. So I signed up for Latin rather than algebra in my ninth grade year. He figured doctors did not need to use much math, except in the investment choices for their discretionary incomes, but Latin might be good for writing prescriptions and naming bones. He had also taken Latin, possibly four years of it.

I ran with a marginally rough crowd in high school. I was still too skinny for any high school sport, but I could draw caricatures and act silly—in fact, I developed into the class clown, or smart-aleck, depending on point of view. I also had blond, slicked back, wavy hair, and I wore cool clothes like pegged cuffs on the pant legs—a bit like Elvis, but I hated him. I preferred rhythm and blues (R&B) or jazz. I was an early fan of the sounds of "colored" musicians. I even attended R&B concerts where I saw live performances of the likes of Fats Domino. I tasted no alcoholic beverage until perhaps late junior year when I drank beer with my pals. No decent girl would drink a beer. This was long before pot was available in Tennessee, and sex with nice girls did not happen. I went to proms and social club parties; I did not study much, accepting the gentleman's B's and C's. In senior year, I buckled down a bit, especially in physics and geometry. I cannot remember reading a single book in high school, except maybe a Mickey Spillane hard-boiled detective (Mike Hammer) story that included some racy action.

During the beginning of my senior year, our new high school physics teacher was delayed from appearing in town for the first few weeks. I was astonished to learn that the school would ask me to "teach" the physics class until he arrived. I had such a grip on chemistry the year before (even though I was a terrible cut-up and prankster) that they thought I might be able to do it. Frankly, I did only a so-so job. But this experience and my favorable performance in geometry class were probably indicators of my future path.

Of the three hundred or so in my graduating class, maybe a quarter started college. This change of educational environment called for a reset: I cut off the long hair and became a square. I followed the path I learned as a responsible store clerk from my beloved Jewish mentors in the retail clothing business. Later when I worked at the University of Missouri-St. Louis, Louie sent his youngest son Paul to study there as an undergraduate, so I had an opportunity to at least in some sense return the favor. The favor I took from Louie was the self-discipline I learned and applied to my job and later in my studies. When I got sloppy arranging or labeling merchandise, he would come over and say, firmly, "A job worth doing is worth doing right." The favor his father, Pop, did for me was to enlarge my horizons so that someday I would think on a larger scale and feel at home in Chicago or New York, or wherever I might land.

I had a few dates in high school but was pretty shy and hardly attractive—skinny, with very crooked teeth. In the spring of my senior year, I met Jane Enneis on a blind date arranged by my friend Jim Corcoran, who had met one of Jane's friends. We all went to the Great Smokies for a Sunday outing, picnicking and hopping stones in an ice-cold mountain stream.

Jane was a year behind me and from an upper-middle-class family across town. She had many friends in the old Southern upper class. In fact, I was her date for her "coming out" dance, which occurred on Chattanooga's Lookout Mountain at a big event called "The Cotton Ball." We rode to Chattanooga with some of her friends, and I had to wear a white dinner jacket. For me it was a wretched experience, but she was able to overlook it and keep me around. We were married three years later in 1959, and we took our honeymoon by driving up to Chicago where I took her to all the spots I so enjoyed as a child with Pop. In our second summer, we drove to New York and watched some shows—I remember "Gypsy" with Ethel Merman. From there we drove on up into New England, passing through Vermont, New Hampshire, stopping in Boston and driving up the rocky Maine Coast. By that time, I was sampling different places for graduate school.

The Mesa Lab Building of the National Center for Atmospheric Research in Boulder, Colorado, with permission.

Harriet Hill, great-grandmother of Jerry, mother of his grandfather, Green Lee Hill.

Jerry's dad Paul "Sanford" North, just enlisted in the Navy, 1944.

Marjorie's older brother and Jerry's uncle, Jefferson "Glenn" Hill, about 1942.

Jerry's parents, Marjorie
and Sanford, long before his
birth in the 1930s.

Jerry's employer and mentor,
Louis Joffe, and his wife,
Sybel.

Marjorie, Jerry's mother.

Marjorie during the war.

The SLOC Store; in Yiddish it means "junk." Jerry worked here Saturdays and all summers during high school.

Jerry, age six.

Jerry and his chickens, about age thirteen.

Knoxville house where Jerry lived during high school.

# CHAPTER 2

# UNIVERSITY AND OAK RIDGE DAYS

The main campus of the University of Tennessee (the real UT, not to be confused with the University of Texas) happened to be in Knoxville (hence, my referring to it as UTK). While requiring some remedial help, I found myself doing well in math and science classes. I was able to support myself by living at home and eventually getting a co-op job in Oak Ridge. Learning how to be a scientist was among my many experiences at the lab.

## University of Tennessee

Since I went to college in my hometown of Knoxville, I could drive my jalopy to and from campus, parking in neighborhoods not too many blocks away from classes. As I entered the university, I signed up for the premed curriculum, mostly because my father urged me in that direction, and I seemed to have enough talent and ambition. My first faculty advisor was a senior professor in the Department of Bacteriology. He had reviewed my package before my first meeting with him, and he told me that I had made something like the 95th percentile in math on my entrance exam. I was only a little above average on language and the other subjects. This math score seemed surprising because I had only taken three years of math in a weak high school, ending with geometry, which I liked immensely. I had not taken trigonometry because I had opted for Latin instead of algebra as a freshman, on my father's advice.

As we began setting up my schedule for the first quarter, my advisor told me that most premed students change their mind before heading to medical school, some even majored in chemistry or (perhaps even bacteriology!). Therefore, it might be a good idea to think about options that would give me flexibility. He had a point. I was only moderately interested in medicine, but I was fascinated with high school chemistry, and we decided a bachelor's degree in chemistry was a good path to take up, at least for starters, since I could use it toward entering med school if that is what I desired later. My advisor also hired me as a petri dish-washer in his lab at 75 cents per hour. I lasted one quarter in that job—deep sinks, very hot water, long rubber gloves: scoop the agar out and wash the little dishes before placing them in the autoclave (scary hot sterilizing oven).

I enrolled in precalculus math, English composition, German, and the special five-credit chemistry course for majors. I also had to suffer through the physical education (PE: calisthenics, running, and general torture) course that was exhausting for me. I was exempted from ROTC because I had a perforated eardrum. The physically grueling combination of physical education, taught by football teaching assistants and occasionally even athletes, and ROTC, taught by tough army guys, was the needless ruination of many of my physically unfit freshman friends. PE was enough for me—thank goodness for my bad ear! After PE class, you had about ten minutes to get showered, dressed, and out the door to your next class, and the core of the UTK campus was accurately named "the Hill," which meant that no matter where you were going after PE, it was uphill. I was so fatigued after one of those classes, that I barely made it to whatever was next. I was pretty frail, and I smoked too much.

My grades for that first quarter were surprisingly good: A's in chemistry and German, B+ in math (trigonometry and precalculus—should have taken trigonometry in high school! It even involved calculating with (my arithmetical nemesis) logarithmic tables with about ten significant figures, unheard of once we entered the age of the hand calculator.), and C in English. The English course was actually a remedial course. Above-average students were able to skip the first quarter, really good ones (like Jane) could skip the second quarter as well. My having to take the catch-up course and my subsequent grade of C indicated my failure to learn how to read and write before entering college, a problem that would surface all through undergraduate and even graduate school. The high school I had attended did not require us to read serious books and learn critical skills in language or literature. Not until senior year did we get any practice or serious criticism of our writing. Part of it, of course, was my own lack of innate talent in those areas. I could not focus when reading, my mind drifting away and reducing speed. It took me decades to get over these handicaps, but the understandable B+ in math and the other grades indicated that I had some native ability, and it gave me hope that I could get through. At the cafeteria, many sat around the circular tables snacking and coffee sipping between classes, working math and chemistry problems. I was a bit faster than most, especially the premeds. It gave me some satisfaction later that some of the slowest science students got into med school, many as legacy applicants (i.e., Dad's a doc). Somehow that path did not sit well with me. In fact, it disgusted me. I was on a strict path to sanctimonious, academic snobbery.

My English instructor was Mrs. Coolie, whose husband was head of the Department of Mathematics. I was again in her class during my second quarter of English composition—I still remember that I read *Out of My Life and Thought* by Albert Schweitzer, and *Return of the Native* by Thomas Hardy, both of which I struggled over. I reread them decades later with great pleasure—in fact, I have read all of Hardy's major novels and several more than once. Mrs. Coolie pulled me aside and told me she had discussed my situation with her husband. She had checked my grades in all of my classes, and she encouraged me to work harder to improve my writing (and reading), that I would surely need that skill later. Maybe I was a "diamond in the rough"? I wish I could thank Mrs. Coolie today. My then-girlfriend Jane corrected my poor spoken grammar constantly (I still said "had went" instead of "had gone," among other atrocities). I eventually got most of those hillbilly scars removed before becoming a teaching assistant in graduate school in Wisconsin where I would have been a laughingstock.

I was a terribly slow reader (about 180 words per minute) unbearable for such an impatient fiend. So Jane (who read at about 300 wpm) and I enrolled during the summer in an evening speed-reading class. I improved to only about 250 wpm and she to about 400 wpm. Now I read many hours every day, but I have never gotten above that very low threshold. I can read a physics book at about the same low speed that I can read a novel or a history book. It does not stop me: I read good books every day. I went to an optometrist who gave me some tests. It turned out I am right-handed but left-eyed. My mom told me after this that my father had stopped me from being a southpaw when I was a baby. Apparently, he did not stop my eyes. I suspect I have a mild form of dyslexia, as do two of my children. I have often wondered how such a mild disability factors later in life. Left-handed persons are right-brained, and that situation is correlated with some good things (leadership, creativity?), and negatively correlated with others (sequential, digital stuff). I tend to be good on the big picture stuff, hopeless and terribly error-prone at the fine sequential scales.

## Co-op Student at Oak Ridge

In the winter quarter, I continued to like general chemistry, and I learned that there was a co-op program that could provide sorely needed assistance for my expenses. At the same time, I would be getting some useful experience for a job later. My interest in med school was slipping away. I qualified for the co-op program, and I was assigned a job at Oak Ridge National Laboratory (ORNL), located in Oak Ridge, which is located only about twenty-five miles from Knoxville. The plan of the co-op program was that the student would alternate from one quarter to the next, first attending classes in school for a quarter, followed by working over the next quarter at the ORNL. My pay was good during work quarters (about $190 a month), and I was able to get into a carpool to Oak Ridge and back to Knoxville every day. My riding companions were nice, older men who were salaried employees at ORNL and charged me only a meager amount for the rides. They picked me up and dropped me off at my home. This great opportunity paid my tuition, which was only $52 per quarter, the next year jumping to $75 per quarter, but still a bargain. I am grateful to the generous taxpayers of Tennessee for this opportunity. My job at ORNL was also enough to support myself and, in the next few years, save enough money to get married.

Obtaining this appointment at ORNL in the Chemistry Division was one of many breaks I encountered in my quest to be a scientist. I worked at the X-10 site, a large campus (comparable to UTK in size) a few miles from the town of Oak Ridge, which was essentially created during WWII for top-secret military purposes. After FBI checks, I received a low-level security clearance, and I wore a badge that registered how much radioactivity I had been exposed to over a period of a few months.

My mentor at ORNL was Dr. Milton "Milt" Lietzke, a research scientist who supervised a laboratory in physical chemistry. Another co-op student and I alternated quarters in Milt's lab. The "section" (like a small academic department) of the Chemistry Division in which Milt worked was dedicated to research that supported several nuclear reactor projects at X-10. In particular, the main focus in Milt's work related to the homogeneous liquid uranium reactor, a pet project of the director of ORNL, Alvin Weinberg.

Instead of the solid fuels used in most reactors, this one used uranium compounds dissolved in strong acid solutions. This high-temperature liquid flowed through pipes and eventually into the nearly spherical volume where it was large enough and concentrated enough to go *critical*—a condition of self-sustained nuclear reactions. At that stage, nuclear fission reactions released enough excess heat that some could be extracted for energy production—at least, that was the theory.

One of my jobs was to measure the solubility of different uranium electrolyte compounds (salts) in pure water (aqueous) or acid solutions at various high temperatures. This kind of knowledge was critical for the operation of such a reactor because an ongoing problem was corrosion of the pipes and pumps in these superheated solutions.

It was in this lab that I came into contact with a graduate student in Milt's group named Richard S. "Dick" Greeley, who worked as a nuclear engineer and group leader for a reactor project at another site in the Oak Ridge complex. Dick was working on his PhD in chemistry at UTK; he used me as a technician in his experiments on the so-called hydrogen electrode for his dissertation work. This was also a high-temperature experiment with aqueous solutions—measurements of great interest in physical chemistry. The experiment involved inserting a small capsule filled with liquid in a titanium shell and slowly raising the temperature in a carefully controlled oven. We put the shell in an oven and turned up the temperature in increments, measuring the voltage between two electrodes (not unlike the two electrodes of a battery) deep in the capsule to see how it varied with temperature. I was acknowledged in three papers published in physical chemistry journals in 1959–60. The acknowledgments read: "We also wish to thank Mr. Gerald North for extensive help in making many of the measurements." Milt Lietzke and Ray Stoughton were coauthors (Greeley et. al 1960). One of these three papers was Milt's most cited article.

I was in awe of all these guys, especially Dick Greeley. He had a bachelor's degree from Harvard (gentleman's C's, he said), after which he went straight into the US Navy to serve in the Korean War. He was the executive officer on a small destroyer. As we worked at some boring task in the lab, I would ask, and he would relate stories from his experiences. To me, Harvard was some mystical place, where only gods pranced around the "Yard." Of course, I was impressed with his Harvard degree, made complete with his Boston accent ("Hahvahd Yahd"), like John F. Kennedy's. He had me over to his home for dinner, and he and his wife actually attended my wedding. They both seemed to me to be the epitome of class, education, and charm.

Dick left ORNL after finishing his PhD and moved to Mitre Corporation where he worked for many years. Mitre is a nonprofit company that provides scientific and engineering services to large enterprises by contract. I met Dick many years later, probably in the 1980s, at a small meeting at the National Center for Atmospheric Research (NCAR). He was representing Mitre on a project proposal related to climate change. We sat around a table with other climate experts, and I proudly told him of my research—I could have burst, knowing he was proud of me. About a decade ago, I saw a letter to the editor of the *New York Times* from him, lamenting his service in the Korean War. He obviously felt guilty about the shelling (potentially killing civilians) from his ship onto the coast of North Korea. In 2013 he published a novel about Vietnam. Dick's scientific career was essentially over after his PhD, though he wrote a number of papers on science and public policy in the journal *Chemical and Engineering News*.

Another of my jobs in the chemistry lab at ORNL was to determine the solubility of different uranium compounds in various kinds of solutions over a range of high temperatures. We had a small oven, like a toaster with a window, that sat on the lab bench. I peered through the thick quartz window, wearing heavy safety glasses, viewing small capsules of aqueous solutions of uranium salts. The small capsules made of quartz were about four inches in length and less than half an inch in diameter. A row of them slowly rotated on a common axis, end over end, their centers fastened to the axel rod. The rotation assured that the solution was well mixed. At low temperatures, I could see the tiny crystal of the compound in the rotating tube. As the temperature in the oven was very slowly changed, equilibrium could be maintained. I could eventually see the crystal disappear into solution and record the temperature of that event in my lab notebook. I generated many tables and plotted up my results.

Each capsule had to have a precise amount of distilled water and a few micrograms of the uranium-salt crystal. The crystal had to be weighed in a clean room (size of a closet and soundproof) with a very precise weighing balance. This was pretty hard for my clumsy hands (remember the mild dyslexia, in addition to a newly discovered lack of hand-eye coordination, [no wonder I could not hit a curve ball]). But the difficult part was sealing the quartz tube with the solution inside. The procedure went as follows. First we cut the quartz tube to a length of approximately four inches with a hack saw and sealed one end using a blowtorch while wearing a hooded helmet with a viewing window and holding the tube with tongs. Now let it cool down. I could do this.

Then it gets more difficult. You pour in your distilled water from a pipette with the previously weighed crystal inside your capsule that is now sealed at one end. Don't spill (this stuff is expensive, a little dangerous, too)! Now hold the little tube with the tongs and dip its closed end into a bath of liquid nitrogen to fast-freeze the liquid, with crystal, inside (you do not want any to evaporate or boil away in the next step because that changes the concentration). The final stage is to quickly move the open tube half full of the frozen stuff to the torch and seal off the top. If you did not do this fast enough, some liquid would not be quite frozen and the tube would explode (splattering the window of your hooded helmet with bad radioactive stuff and broken quartz). When this catastrophe happens, you have to start all over. Milt could execute this entire procedure with the greatest of dexterity and ease, never failing. I must have tried it a dozen times and never was able to do it. Well, maybe I did once by pure accident.

Milt was incredibly patient with me, but in the end he had to close all of the capsules for me; once I got the tube cut and closed at one end, I measured out the solvent (the liquid that does the dissolving) and weighed the solute (the tiny crystal being dissolved) and inserted them into to tube ready for sealing. I would run across the hall to his office and hope he was there. Those weeks were important for my career and life. I was beginning to realize that I was not a very good high-precision worker in a lab when the job called for great dexterity. I knew then that I made the right decision not to be a surgeon!

As each of my academic quarters finished, I could not wait to get back to my job in the Chemistry Division. The scientists I worked with were kind and everyone was on a first-name basis. There was Milt and Joe

(Joseph Halperin), his office mate. Milt had a PhD from Wisconsin, Joe had a PhD from Chicago, and the group leader, Ray Stoughton, received his PhD under the direction of Nobel Laureate Glen Seaborg from the University of California (UC), Berkeley. Ray had a bad leg from polio when he was a child, but he never stopped moving. Being around such educated scientists made me feel that I wanted to be like them, independent researchers all.

To enjoy a middle-class life, fueled by work that one loved! Well, maybe not if it involved quartz blowing combined with liquid nitrogen. They looked down on the lowly professors at UTK (ironically, Milt eventually moved over to UTK as a professor). I began to see how the prestige of scientists fit into a hierarchy based on their skills, credentials (pedigrees), and their productivity. There was even a pecking order in the Chemistry Division.

The group welcomed me for lunch in Ray's office every day. Others joined our lunches occasionally, but I was one of the regular guys. We all brought our lunches (mine a bologna sandwich). They each had peculiar outside interests, but I especially remember an interest in parapsychology and other mysterious weird subjects. At that time there was a Prof. Rhine at Duke University who studied paranormal phenomena, doing experiments and applying statistical methods to the results. Of course, none of the experiments ever actually demonstrated anything, but my friends continued to search the literature on the subject. Books were read and discussed. I thought it demonstrated that they were very open-minded. I became interested, but not for long—too nutty.

Milt was one of the most interesting and unusual men I have ever known. Aside from his expertise and skills as a chemist, he could read and speak many languages, including German, French, Latin, and Greek. He had read the New Testament in its original Greek, recorded almost two thousand years ago. He had memorized numerous opera librettos, and he could sing them. He loved number theory and would speak about it in the office during coffee breaks, putting proofs on the blackboard. Milt was also a master of limericks, and he could rattle them off by the dozens. (I had never heard one before.) He also had raunchy rhymes for every letter of the alphabet.

Having seen what a klutz I was in the lab, he taught me how to use the computer housed in our same building. Milt's wife was a programmer, and he admired her tremendously. He acknowledged her for computations and me for making the measurements in a couple of his papers.

The computer at ORNL was called the Oak Ridge Automatic Computer and Logical Engine (ORACLE). The acronym matches the name of the place in Greek mythology where men went to query about the future. The ORACLE was installed at ORNL in 1954. It was one of the first operational digital computers. I learned to write out a program in the hexadecimal number system. Hexadecimal is based on sixteen as opposed to our usual ten based (decimal) system. You can indicate a single digit number as 1, 2, ..., 9, A, B, C, D, E, F.

On the paper tape, you can indicate the digit by a series of four spaces. For example, unity is represented as blank, blank, blank, punch. A punch is called a *bit*. A hexadecimal digit (called a *nibble*) is composed of four bits. Sixteen (or F) is represented by punch, punch, punch, punch. So a sequence of numbers can be punched into the tape. As the computer takes in the tape, it advances by one of these steps at a time. You can also code instructions, for example, to multiply two numbers located in specified locations and store them at some other address. There were about two thousand addresses included in your code. These coded instructions could be strung together, and there were possibilities for loops, logical *if statements*, and so on. It takes a while, but one gets used to the hexadecimal language and can read what's being transmitted on a strip of paper tape.

I would go into the key-punch room (incredibly noisy, well over one hundred decibels, roughly the sound level in a New York subway station) and begin punching up my program. If you make a typo, you can rip out the tape, duplicate it until you come to the error. All this requires the ability to read those hexadecimal patterns in the punch holes. You could always ask someone in the room for help, but they were all very focused on their own work and treated me like an idiot. I was able to manage, but I am a pretty bad typist (I could not write a book today without a word processor), making the work pretty long and grueling. But I did learn to do calculations that were pretty complicated. This would help me through the transition to my next job at ORNL.

I went nonstop during my quarters at the Chemistry Division. While I was working at ORNL, UTK offered evening courses in the town of Oak Ridge for employees trying to better themselves. In particular, they offered calculus (sophomore level with some calculus as a prerequisite) and university physics, both courses suitable for engineering and science students. One quarter I took both in the evening, a five-hour calculus course, and a four-hour physics course. Needless to say, I was busy every night and weekend. I stayed in a dorm in Oak Ridge that quarter. But I took night courses in either Knoxville or Oak Ridge every quarter while working. I could take a bus to work and a bus to Knoxville on Friday night. I never seemed to stop. I weighed about 125 pounds at 5'9" height. I smoked too much (it was the 1950s), and thought eating was a waste of time—I wished I could just take a pill instead of making lunch and then taking (wasting) the time to eat it.

## Married Life

Jane Enneis and I were married in July 1959. Her father was a doctor who served as the Public Health Commissioner of the Knoxville area. I only met him once. By that time he had lateral sclerosis and died soon after. Jane's family was originally from a small town in Georgia. They were definitely upper-middle to lower-upper class in the Southern aristocracy, as were most families of doctors at the time. Her father did not make as much money as the other doctors because he worked for the city. Besides his MD degree (I do not know where) he had an MS in public health from Harvard. He was a hero who took on local businesses—most publicized were the grocers—for their unsafe practices, not to mention putting thumbs on the scales.

There were some adjustments for me. At first, her mother was rather suspicious of me since I was grossly unpolished. But as we got to know each other, she came to like me as she found that I was ambitious and would provide well for her daughter. I dated Jane starting in her senior year in high school, as mentioned earlier. She became a psychology major at UTK and was one year behind me. She was smart and enjoyed reading. She could read French with ease. I became a much more liberally educated person because of her influence.

After the wedding, we took a furnished apartment ($45 a month) on Clinch Street only a few blocks away from the UTK campus. The apartment was not a long walk to the location of the childhood home of novelist and screenwriter James Agee, whose masterpiece was A Death in the

*Family*, a novel set about a generation before my time but still reminiscent of my days in Knoxville. Agee's father had died in an alcohol-related, one-car accident; my father was in a similar accident but it happened in 1956 when I was still in high school. My father lived, but his injuries and habits destroyed him as a man, and he died at age fifty-eight in 1968.

Jane and I took many weekend rides from Knoxville to the Great Smoky Mountains National Park where we took long hikes. We walked up Mt. LeConte along several different trails. The most popular for Mt. Leconte is the Alum Cave trail, which is 11 miles round trip starting at elevation 2763 ft. up to 6593 ft. Another is the Rainbow Falls trail, which is 13.8 miles round trip. The Bullhead trail is 14.4 miles and starts at 3993 ft.

Some favorite peaks along the continental divide include Gregory's Bald—the trail is 11.3 miles round trip. Others we hiked include Siler's Bald, the Chimneys, and Andrew's Bald. We always took a lunch along and ate at the summit. These were hikes, not climbs, but round trip meant getting started early so as to get back to the car before dark. We would occasionally see a bear on the trail, but they were always shy and scampered away at the sight of us.

Both Jane and I were plenty busy in school and at work. She was in a sorority, we had fun with friends, and we went to movies when we could. She was very kind to put up with my long hours at work, commuting, and in classes.

## Losing the Taste for Chemistry

In my sophomore year, I took chemistry courses that involved labs. As already was evident, I lacked lab skills and the required patience. I always rushed to get the experiment or the laboratory analysis finished, so I could get the hell out of there. Along about this time I took an advanced physics course and advanced calculus (to the dismay of my chemistry advisor), both of which better suited me. I then took physical chemistry, using the famous book by Daniels and Alberty (I still have it). Aesthetically it contrasted so much with the elegant math and physics courses that I could not bear it. Moreover, the chemistry professor was terrible. I cannot even remember his name. He had students stand at the blackboard doing homework problems. I found this practice to be excruciating. I hated listening to a student discuss his attempt at a solution and the professor, trying to get him through it (or was he?). I always thought the left side of my stomach was eating the right side during these ordeals of the belabored students, sometimes including me. Frankly, it was a lazy way to teach. Most of physical chemistry revolves around the science of thermodynamics. Years later I came to love thermodynamics as taught by physics books and even coauthored a book with my good friend Dr. Tatiana Erukhimova on atmospheric thermodynamics while at Texas A&M—just one of those funny ironies (North and Erukhimova 2009).

The wise Herr Prof. Sommerfeld said: "Thermodynamics is a funny subject. The first time you go through it, you don't understand it at all. The second time you go through it, you think you understand it, except for one or two points. The third time you go through it, you know you don't understand it, but by that time you are so used to the subject, it doesn't bother you anymore."

## Transition to Physics

I loved math and physics, and I began to dislike chemistry. Unfortunately, a co-op program for physics majors did not exist. It was available at UTK for engineers and chemists, but not physicists. My chemistry friends (especially Dick Greeley) went to work to find me a job at ORNL that would be physics-related. This turned out to be in the Thermonuclear Division, located at the Y-12 site. During the war, Y-12 was the main location for uranium isotope separation using "calutrons." At this time, Oak Ridge was more famous for its gaseous diffusion process that was used to separate isotopes of uranium. You can read about it at energy.gov/management/k-25-gaseous-diffusion-process-building.

## Direct Current Experiments

The Thermonuclear Division worked on projects related to the prospects for "fusion" nuclear energy. This nuclear mechanism utilizes the combining (fusion) of light nuclei such as $^1H$ (a proton) with another proton or perhaps with a $^2H$ (deuterium, whose nucleus is composed of a proton and a neutron) to form a Helium nucleus. This is the thermonuclear principle used in the H-bomb. But in this case, as with all the other nuclear projects at Oak Ridge, the aim was to tame these processes for peaceful applications. Hence, we were trying to build a fusion reactor or at least conduct theoretical and experimental studies on how this might be accomplished. The group to which I was attached worked on an experiment called the Direct Current Experiment (DCX).

Another wonderful experience was in store for me. I first managed to arrange things so that I worked only thirty-two hours per week. This way I could live in Knoxville and take a course in the morning and then drive to work. I could also take UTK night courses at either the Oak Ridge (math and physics) or the Knoxville campus (economics, history, and other requirements for my BS degree). UT was a very old-fashioned Southern university with a classics-oriented curriculum. There were not many BS degrees, most were BA degrees where there were requirements such as three years of a foreign language (including a serious third-year literature sequence). For example, Jane was working on a BA with a major in psychology, and she took a third year of French literature. I might have liked a third year of German literature, but it never would have fit my schedule. Luckily for me the Department of Physics had just established a BS in physics curriculum, whose main advantage for me was the elimination of that third-year requirement in literature. Of course, we still had a full year of English language literature, loaded with poetry as well as short and long pieces of fiction. I also took a required full year of world history and two full years of social science. I was to lose some credits that I had taken in chemistry, except that I was able to substitute some of the chemistry courses for the physics thermodynamics course and a full-year course in biology (either zoology or botany, not half and half of each). This latter hole in my physics background had some repercussions later because biology is important in climate science. Finally, I thank UTK for pushing me to take courses in literature, philosophy, and music. These paid off later in life as I love to read literature, history, and philosophy. I suppose most of these diverse requirements at UTK and other Southern public universities have melted away over the years as the market for students has shifted the business model of higher education (more venting about that later).

Yet another fortunate circumstance awaited me in the Thermonuclear Division. Security was tighter at Y-12 because of its association with isotope separation, but I had nothing to do with those matters. I was primarily hired as a computer programmer. This is somewhat laughable today as I am a terrible programmer. But in those days, I was better at it, and only very few scientists were skilled at it. I was assigned to assist Ms. Mozelle Rankin, a mathematician who was an expert on ORACLE programming as well as other machines that were emerging at that time, such as the IBM 704 that had just been installed at the K-25 site (ten miles away). The programming language for the IBM 704 was FORTRAN, a much more user-friendly computer language than the machine code in hexadecimal characters on paper tape. Instead, FORTRAN code was written on punched cards.

Mozelle was absolutely wonderful, a fine applied mathematician, numerical analyst, and friendly, patient mentor. Her office, which I shared with her, was right next door to the leaders of the division, Dr. Art Snell (our division head), Dr. Ed Shipley, and Dr. P. R. Bell. It was close enough that I could hear them arguing about science and engineering problems related to our project. Of course, I got to know them because I was writing programs for Shipley and Bell— although Snell was a PhD atomic physicist, he was too busy with the administration to have much contact with me.

Shipley was a lively, even flamboyant, PhD in electrical engineering and former professor, who would call me into his office and dazzle me at the blackboard with sketches of his ideas. It was clear that he had been a teacher and he missed it. He was very creative, persuasive, and lovable. P. R. Bell was fascinating. He was brilliant, argumentative, and ever in a hurry, wanting answers from the calculations he had me doing. He always wore a three-piece suit and colorful tie. He was nearly blind from the work he did during the war. He had worked at the Lincoln Labs at MIT. X-Rays (or even worse, radiation) permanently damaged his eyes. Because of this, he wore some tiny lenses attached to the rim of his glasses that he could fold down in front of his regular lenses, similar to the arrangement used by jewelers. He clearly liked me, but he wanted his answers quickly, and they had to be correct. He would look at the output of my programs, flicking down that lens, and tell me my result could not possibly be right. His physical intuition was extraordinary. He could not do math anymore but that intuition and sound scientific judgment worked remarkably well. Often he was right, but not always. When everything worked, he patted me on the back and gleefully squeaked; then he ran out to tell everybody he could identify. His manner put off some of his peers. Years after I left ORNL, he married Mozelle.

There was a man in the division named John Luce who had started as a technician, working on carbon arcs. These arcs are used in big search-lights to provide a very bright point source of light in front of a curved mirror to focus a strong beam into the night air. They were used during the war to scan the night sky for enemy aircraft, but now only for gala events and publicity. Oak Ridge became interested in these arcs for other purposes. One was the possibility of disassociating hydrogen molecular ions ($H_2^+$) into their atomic constituents: $H^+$ and H. Luce came up with the idea that we could create a magnetic field and run the arc perpendicu-lar (or across) to the field lines. He tried this and it worked. He found that it could disassociate molecules into their atomic constituents. This effect could be used to create a plasma.

A plasma is a high temperature gas that is so hot that some of the electrons are no longer tightly attached to the atoms. If one could create such a plasma at high enough temperatures, there is the (remote) possi-bility of having the atomic (actually nuclear) collisions be hard enough to induce fusion reactions leading to Helium nuclei and the release of huge amounts of energy. The Thermonuclear Division emerged at Oak Ridge, and resources were allotted to see if a Luce-type machine might be built that could harness the H-bomb concept for peaceful purposes.

It was this creative idea that developed into a device called DCX, the Direct Current Experiment. When I arrived, DCX was up and running. A beam of $H_2^+$ ions was to run through the arc producing the plasma. The atomic ions ($H^+$) then were confined by the magnetic field in the cham-ber. The magnetic field was very strong and it was pinched such that its lines were squeezed together at the ends of the chamber. This pinching of the magnetic lines formed a kind of mirror reflecting ions back toward the interior of the device. The longer it ran, the denser the plasma would become, and it was guaranteed to be hot because of the high speed of the newly generated $H^+$ ions. The configuration was dubbed a "magnetic bottle."

Theory suggested that this should produce thermonuclear reactions in the interior if there was not too much leakage through the mirror and out of the machine. Mozelle and I wrote codes that calculated the trajecto-ries of the ions. The problem always was that if the ion's helical trajectory was tilted too much along the magnetic lines, it would fly right through the mirrors. We were able to show that the magnetic bottle worked for a large class of the orbits. Unfortunately, mild collisions of the ions with each other would inevitably cause some to eventually be oriented into the angles that would cause them to leak. DCX failed because it could never achieve enough density of the ions to produce fusion reactions.

## DCX-2 and the Physicists

But there was hope. The next step was to build a bigger and better device called DCX-2. I participated in the design primarily through calculations related to ion trajectories and the magnetic fields. Some new scientists soon entered into my immediate proximity. These were real PhD-level plasma physicists. There were three that made a strong impression on me: Al Simon (head of the theoretical group), T. Kenneth Fowler, and Robert L. "Bob" Mackin. Another plasma physicist was Edward Harris, a physics professor at UTK, under whom I had taken a yearlong course on modern physics. Harris was a wonderful teacher, and he was also a consultant to the division. At one point I did some calculations for Harris for a famous paper he wrote on plasma stability.

Ken Fowler worked with Mozelle, and he occasionally had a brown bag lunch with me. He had a PhD in theoretical physics from the University of Wisconsin. I got to know him well, and he provided me with advice that eventually led me to his alma mater (also Milt Lictzkc's) for my own PhD. His dissertation was in high-energy theoretical physics sometimes known as elementary particle theory. He worked under Kenneth Watson, who moved on to UC Berkeley before I arrived at Madison. Ken suggested that in graduate school I should study elementary particle theory because it was at the frontier of physics. The leading edge is where you must use the very most modern mathematics. After you get your PhD, you can change fields, as he did into plasma physics, for example. He also told me that Wisconsin was just right for me since he knew my strengths and limitations very well. I think he compared me to himself (actually that would have been very generous of him). He also told me that at Wisconsin they have a reputation for "taking care of their students." The treatment could be very brutal at some schools. A few years later, I did go to Wisconsin. This calls for another quote from my academic great-grandpa

Sommerfeld: "If you want to be a physicist, you must do three things—first, study mathematics, second, study more mathematics, and third, do the same."

## My First Publications

I calculated magnetic fields for various wiring-coil designs as part of the DCX-2 project in the Thermonuclear Division. Introductory physics books show that an infinitely long cylindrical metallic shell with a uniform electrical current density going around it produces a magnetic field that is perfectly uniform in the interior, and the magnetic lines of force are parallel to the axis of the cylindrical shell. The formula for the magnetic field strength is simple, and it provides examples in all elementary physics books. This remarkable property is the basis for DCX (or Luce-type) machines. If the hollow conducting cylinder has only a finite length, the field will be nearly uniform near its center, lengthwise and along the axis, but diverging away from the axis at the ends into the space outside the tube.

If ions are injected into a cavity like this, they will follow helical trajectories circling the magnetic lines of force. By increasing the electrical current density near the ends of the tube you can make the magnetic lines pinch together before they diverge outside the cylinder's ends. The pinched lines (sometimes called tubes of flux) serve as a "reflector" of ions traveling toward the ends of the cylinder. In DCX-2, the coil supplying the current surrounding the cylindrical cavity was to be in loops rather than smoothly along the cylindrical shell.

I considered the problem of an infinitely long cylinder consisting of equidistantly spaced discrete loops (wires) of current. I found an analytical solution to the problem that showed that along the axis of the "lumped" circular currents there would be a ripple whose strength was evident by a simple formula, based on the spacing, radius, and other characteristics of the loops. In finding the solution, I used a Fourier series technique, and in evaluating the coefficients of the series, I found a really interesting trick that allowed me to solve the problem for the ripple's strength. I was also able to find a mathematical means of extending the solution off the axis of the cylindrical geometry. I was so excited that I had found this solution, my very first discovery in physics (well, actually engineering). All of my close mentors were happy for me (Rankin, Fowler, and Mackin). I wrote a Technical Memo, "Some Properties of Infinite Lumped Solenoids," ORNL-2975, by G. R. North, August 1960, that was distributed widely in the division and some copies across ORNL. I made sure my advanced calculus professor, Dr. Walter Gautschi, was on the list.

Soon after the publication was distributed, Dr. Gautschi called and wanted to see me. When we met, he congratulated me but had to inform me that the fancy trick I discovered was known in Europe as the *Poisson Summation Rule*. He referred me to a calculus book written in German by Alexander M. Ostrowski, who I just found out was actually Gautschi's doctoral advisor—such a modest man, Gautschi never told me of it. I do not think I have ever seen it in an English language book. I found the book in the ORNL library. I felt bad that I was not the original discoverer. A legendary mathematician had scooped me by about two hundred years. Of course, Poisson did not have the magnetic field application in mind for use of his formula. Many years later, I wrote the solution up and published it in the *American Journal of Physics*, acknowledging both Gautschi and Poisson (North 1971).

Soon after, my PhD colleagues invited me to be a coauthor on two papers to be published in the journal *Nuclear Fusion* that came out in 1962, my first papers to be published in a refereed journal. Dr. Wilhelm F. Gauster, one of the coauthors, followed my work on the magnetic field calculations. He was a German immigrant who came to the United States after WWII. He was a towering man, bald with olive skin, probably in his fifties at the time. His expertise was in magnetics, a specialty in applied physics. He worked out the optimal currents to be applied in the coils of DCX-2 to maximize the uniformity of the field. In the optimization process, he used an eigenvalue method that I found to be especially elegant. He wanted me to stay at Oak Ridge and be his PhD student, but by that time, I was already thinking of Wisconsin. I was going to be an elementary particles guy, not a magnetics guy.

DCX-2 finally was assembled and turned on. What would happen? As when the first hydrogen bomb was exploded in the Pacific, there was speculation that the entire atmosphere of the Earth might be ignited. That did not happen, but there was to be a tragic accident. The "wires" that went into coils of DCX-2 consisted of copper tubes of about a one-inch thickness and a square cross-section covered with insulation, but the tubes were hollow so that cooling water could be circulated while the coils were carrying the current. Scientists were standing about during the big event, including the chief engineering physicist for the project. The magnetic forces of the coils were so large that one of the coil wires blew apart and like an uncoiling snake struck this man and killed him. Everyone was mortified by this tragedy, especially since the poor victim was such a nice young man. There was one other fatality in the division while I was at ORNL. An electrocution occurred in a physics laboratory during an experiment. Again, the victim was a nice man that I had known.

I received my BS degree in physics the very month that my memo was published (August 1960). I had managed to get my degree in four years while the co-op program was expected to take five. Instead, I opted for the 32-hour week with night classes. Jane was a year behind me, so I stayed on in the Thermonuclear Division for another year. During that year, I was to take two evening graduate courses in physics at UTK facilities in the city of Oak Ridge. One was a mathematical physics course taught by Al Simon, one of my mentors in the Thermonuclear Division.

DCX-2 also failed to deliver the confinement of a plasma of high enough density and for sufficient duration to make fusion reactions or even neutrons from the nuclear collision of deuterium ions when the heavy hydrogen isotope was used in the ion beams. I think this might have ended the serious efforts at Oak Ridge for a fusion program. Mozelle and Bell stayed on at the lab, while many of the other young physicists moved on. Bob Mackin moved to the Jet Propulsion Lab (JPL) and eventually became an executive there. His knowledge of plasmas applied nicely to the Earth's outer atmosphere, which was of interest to JPL. I did get into email contact with him many years later since I had a project at JPL, but the schedules of my visits there never worked out for us to meet. He did tell me that he followed my publications in *Science* with Tom Crowley, my paleoclimatology partner of the 1980s and 1990s. He expressed curiosity that the papers had no math in them. It was unusual because most of my papers have probably had too much math in them. I will mention Ken Fowler in the next chapter.

In the summer of 1961, we left for the great unknown in Madison, Wisconsin, where Jane was to be admitted as a graduate student in psychology.

Jerry, high school senior photograph, 1956.

The computer ORACLE in 1954 at Oak Ridge National Laboratory, Oak Ridge, Tennessee. Jerry was a junior physicist in the Thermonuclear Division.

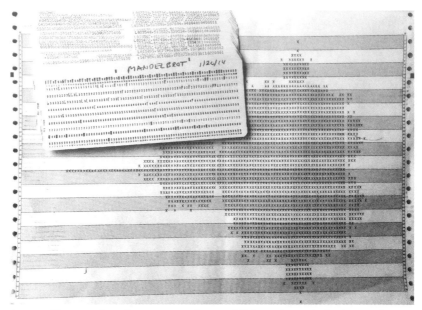

A page of input cards and output paper from an IBM 704 (or equivalent) as used at ORNL when Jerry worked there. With permission from the blog of Ken Sherriff.

Jerry (left) and Charles Goebel (right), his dissertation advisor at the University of Wisconsin, Madison. Photo taken in 2006 during a colloquium visit to Madison.

# CHAPTER 3

# THE LOVE OF PHYSICS AND MORE

My transition to physics was successful. I went on to graduate school for my PhD with a dissertation in quantum field theory. It was good enough to get a postdoc job at the University of Pennsylvania. The job market was contracting and the competition for jobs intensified. I took the only job I was offered.

## Graduate School

From school to school, undergraduate physics curricula are roughly the same. The sophomore course that requires calculus as a co- or prerequisite is about the hardest course you can take. All of the concepts are new and difficult to grasp. What are mass and force? Acceleration is easy: even in vector form, it is just the rate of change of velocity. Even so, the basic rules of mechanics are difficult to incorporate into the intuition. No wonder it took so many centuries for this to become clear, culminating in the work of Galileo, with finishing touches by Newton. But once you get it, it sticks. You can suddenly solve problems like predicting where a cannonball will fall when you know the initial elevation angle of projection and its speed. Such trajectories can actually be calculated with pencil and paper! There is an analytical solution to the ball's trajectory. It is just a parabola, which you just learned about in calculus. This elegance and the quantitative results beat chemistry for aesthetics every time.

Of course, higher-level physics is not for everybody. It gets harder as you ascend this ladder. But like hiking up the most magnificent mountain, the higher you get the farther you can see—you begin to get the big picture. A physics education has many advantages. There are all those steps along the academic obstacle course in which you learn a great many problem-solving skills, such as first stripping a problem to its bare bones, solving that, and then working your way up, step by step, to the problem that approaches all the complications of the real world. For example, that cannonball calculation might need corrections for air resistance and wind. There is probably a career fit for you at one of those levels, BS, MS, PhD, and postdoc, where you have gone as far as your abilities and uncontrollable personal constraints allow. I say constraints because some people can work longer hours while staying focused and others less so. The workaholic index varies from one person to another. Some can take years of long-term uncertainty as postdocs drifting from one perch to another for years, others have acquired families and seek some form of job security. Many factors determine the niche that works.

Students get good laboratory exposure in the physics curriculum, learning to wire up instruments, pump out vacuums, measure vanishingly low pressures, look for leaks, create ion beams, and much more. Some even learn to weld pipes together along with other machine shop skills. They also learn to program computers because so many problems in the real or engineering world are too complicated for those elegant analytical solutions. Elegance does not only come from analytical solutions to math problems, but it is equally aesthetic to design a neat experiment in the lab or on the computer.

Sadly for me, I was an impatient klutz in the lab and an impatient and flawed programmer. As mentioned before, the problem is some sort of dyslexia that makes my fingers go one way when I wish they would go another. So I am a physics guy but with a rather narrow range of work skills. Lots of men and women have become famous with only these skills, but the competition is keen among them, and they do not seem to be of much use in the private sector unless they are extremely good. Some of my grown-up acquaintances at Oak Ridge who thought highly of me asked, "Why would you want to be a theorist? Most are not good enough to do anything meaningful even though they have a PhD, and they wind up writing long complicated computer programs." But dumb as it was, I wanted to be like the theorists I knew and read about, but I was not smart enough to be like the giants or even the hordes of semi-famous theoreticians.

Another benefit of the physics education, not often mentioned, is that we can understand physics well above the level at which we can gainfully practice. We may not ever break new ground and startle the world of our peers at those levels just above our comfort zone. This means there are aesthetic experiences open to us through reading at a level above our own level of practice. It might be something like being able to read a foreign language (with a dictionary) but not able to tell or get a good joke in that language. Many of us can read extensively in the history of physics, especially the biographies of the greats. This adulation makes us feel that we are part of a greater and very special society. We express it with arrogance toward just about everyone. I was into that.

In graduate school, I made good grades in my theoretical physics and mathematics courses. There were colleagues around me who were smarter or quicker, but I harbored the illusion that I was wiser. My high point came in my third year in Madison when we had to take our preliminary exam. The exam was scheduled for mid-March. My cohorts and I studied for three months straight for this exam. My mind was on it 24/7. There was a physics book or pile of notes on every flat surface in the apartment, including the bathroom. The exam consisted of two or three days of written questions followed a week later with a two- or three-hour oral exam. One of my best friends (with a great career ahead of him) had to go out during the written exam a couple of times to vomit, then return. My advisor was a high-energy theoretical physicist. I did not care much for the experimental side of that subject, even reading about it (in those days, now I have a greater appreciation for it). This alone probably said that I could not be a grown-up player in high-energy theory because the real players loved both. There were so many elementary particles (pi-mesons, K-mesons, neutrinos, protons, neutrons, etc.) that it all sounded like chemistry to me. I figured I would never work in that particular subfield anyway. I wondered whether, if I failed the prelim, I might look for a weaker school and try for a PhD in mathematics.

One of my physics mentors, Ken Fowler at Oak Ridge, said I should do my dissertation in high-energy theory then go into something easier, or newer and developing, when I got out. He did that successfully, going into plasma physics (the study of ionized gases, with application to peaceful nuclear fusion research, also applicable in the Earth's outer atmosphere). I knew something about plasma physics because of my background at Oak Ridge, so in January, I began sitting in on the plasma theory course taught by Hal Lewis. Sure enough, on the written part of the prelim exam, I had to choose three long problems to work out. I chose atomic physics, plasma physics, and something else that I cannot remember. Fresh out of sitting in on Lewis's course, I found myself doing the plasma physics problem that took ten pages of work to get through. Actually, I took three courses from Lewis, special and general relativity, applied theoretical physics, and the plasma course. He was a marvelous lecturer, never a textbook, all the lectures were made of original material, always sporting his white short-sleeve, open-collar polo shirt. He gave no exams, expecting students to be mature enough to do as they please, taking from him what he did best. I still have his lecture notes, which have the quality of a good textbook.

I took a full-year graduate-level mathematical method physics course (credit from UT) from Albert Simon at Oak Ridge during my extra year in the Thermonuclear Division. Simon was a theorist who was the leader of the theoretical group in our division. After I left for graduate school, he left for a faculty position at the University of Rochester in mechanical engineering. Because I had this course, I did not take a similar course at Madison; instead, I took the graduate complex variables course from the great mathematician and excellent teacher, Walter Rudin. Partway through the first semester, I was one of only two physics majors (the other guy fell by the wayside in a few weeks) in the course; the rest were first- or second-year graduate students in mathematics (see Rudin's fascinating memoir: Rudin 1997).

Partway through the first semester, I was having minor troubles (set theory and topology, for just two examples) with the course due to my spotty background, so I went in to speak to Rudin. I asked him whether there was any book that I might consult to bring me up to speed on the untaken prerequisites. I had taken an advanced calculus course at Oak Ridge under Walter Gautschi, a well-respected numerical methods expert, who later became a professor at Purdue. Rudin suggested that I might look at *his own book* on real analysis. I later felt pretty stupid, because it is a classic, perhaps the most widely adopted book on the subject for senior undergrad math majors. I still own a copy of that book, and I recently perused the opening chapters for fun. I took the second semester of Ruden's course but dropped it halfway through because I had satisfied my math requirement, and I was getting too busy with physics and study for the prelim exam. I also got A's in graduate math courses in linear algebra and group theory, both without the required undergraduate prerequisites—and, of course, I took them in the wrong order. Punishment seemed to suit me.

The oral prelim exam for me seemed horrible. My mind flickered, dimmed, and occasionally it went blank. I walked over to the window and stared out for seconds (it seemed like minutes) at a time. The physics examiners like to get you onto a silly question such as, "How does a sailboat work? What if the water is an ideal (frictionless) fluid, could you row such a boat? What if it is in molasses? How does the sail work to make you go forward when the wind is in your face?" Me, I am scared of water. My score on the oral was 77.

I felt that I had done well on the written exam, however. When the scores came in my best friend Phil James (still in touch today, also still about eight times smarter than I) and I were comparably performing young theorists, and of course we were competitive with each other. There were about twenty students taking the exam, and by tradition, the student with the highest score was to manage the spring picnic—a much sought-after honor. Phil scored a few points above me and I was second best. I started looking through the graded exam papers and found that on the plasma physics question I had scored one point out of a possible ten. That had to be an error because I knew that subject and that problem! The grader had given me a score of 1 out of 10. The person who compiled the grades mistakenly took the first *page number* of my stack of pages to be the score. Correction of this grading error was just enough to put me a negligible smidge ahead of Phil. We shared the same advisor, high-energy theoretician Adam Bincer, and he suggested to me that we call it a tie, since Phil had already been declared the winner. I agreed completely. Of course, some faculty members sniped to Bincer that I had chosen the easy atomic physics question instead of taking on the particle physics question. I simply chose the easier question out of fear, actually panic. Bincer went away to Brazil on sabbatical for a year when Phil and I had not yet really gotten into research. I switched to Charles Goebel, and Phil switched to Marvin Ebel as our new mentors.

I want to say a few more words about Hal Lewis. He had been a student of J. Robert Oppenheimer at UC Berkeley. After his MS degree, he served in the US Navy during WWII as an electronics wizard. I sometimes wonder if he had been in (or perhaps taught) some of the Navy classes on radar with my father at the Naval Station at Great Lakes, located near

Chicago, during the last years of the war. After the war, Lewis continued working as a graduate student under Oppenheimer, who was then director of the Institute for Advanced Studies in Princeton. Technically, Lewis received his PhD from Berkeley.

After about five years as a research scientist at Bell Labs, Lewis started his academic career at an assistant professor at Berkeley but resigned when the state of California forced all state employees to sign a loyalty oath during the days of McCarthyism. He moved to the University of Wisconsin at that point. In the 1960s, he moved to UC Santa Barbara to found a center for theoretical physics, taking two theorists from Madison with him. Lewis had few graduate students. He had already begun heavy consulting work for the government, especially the military. He became the chairman of the JASON group for a number of years and devoted much of his later career to defense research. I bring this up because I had occasion to give some lectures and had some other roles regarding climate modeling with the JASONs myself, many years after graduate school. Lewis later had opinions on climate change that I will further explore later.

A thorny obstacle in getting a PhD in those days was the foreign language requirement. At that time, every PhD program in the sciences required reading knowledge of two foreign languages. At Madison you could only use French, German, or Russian. I chose the first two. I had taken two years of German as an undergrad, but I was rusty. I spent a good part of a summer preparing to take the exam. The German exam was to be based upon a book, which you could choose long before you took the exam. On the day of the exam, you were to bring your book to the examiner and he would open the book to a random passage (one with no equations that could help you). You were to interpret the passage, perhaps one page, orally. All this was conducted in front of a room full of others nervously waiting for their turn. I thought back to that physical chemistry class. The German examiners were rumored to be very grouchy. I interpreted my passage from a German quantum mechanics book. I thought I did pretty well, certainly, I understood the material. The examiner said, "That was lousy, but I will pass you."

I got out of there as fast as I could. The next summer I turned to French. The book selection was up to the student; I chose a quantum mechanics book by quantum pioneer and Nobel Laureate Louie DeBroglie. On the French exam, we were to write out our translation. Naturally, French writing is easier for Americans, and I only had to study about four weeks to pass with flying colors. My high school Latin helped—what irony: I took it at my father's advice for studying medicine. I bring this ancient relic of a requirement up here because it really took a lot of time (about two summers of study), and of course, everyone was becoming aware that English was the new *lingua Franca* in science. Some of my friends had to take these exams over and over before they could pass them and get on with their PhD research. In those days, engineering students did not take a foreign language as undergrads, and of course, the popular choice of Spanish was not recognized. The requirement was abolished throughout American universities a few years later. While the practice seemed cruel at the time, I still enjoy my smattering of French and German as it pops up often in my reading of history and literature.

I think the tie for the prelim championship was my high point as a high-energy theoretical physicist. I managed to get a dissertation done, and to my surprise, I secured a postdoc position in the prestigious Department of Physics at the University of Pennsylvania. My pal, Phil James, nabbed one at Illinois to work under John David Jackson, the author of the widely adopted graduate-level electricity and magnetism textbook. The first edition of Jackson's book appeared while I happened to be a student in the course at the University of Wisconsin (I was proud to make an A in that course!). Jackson's third edition is still widely in use. Somehow Goebel wrote a strong recommendation letter for me, although I always thought that he regarded me as an idiot (which, compared to him, I was).

A fraction of my cohort of students got postdocs, and a few experimentalists got academic jobs at big-name schools, but later were declined for tenure. Two high-energy experimentalists got faculty positions at Madison; both were very successful, one eventually served as department chair and the other later moved to UCLA for the rest of his career. Perhaps the most successful was a solid-state experimental physicist, a long-time friend and bridge player within our circle, Paul Peercy (the guy who took vomit-breaks on the prelim exam). Paul received and made the best of an extraordinary postdoc at Bell Labs and later a number of increasingly higher-level jobs, eventually returning to become Dean of Engineering at Madison. Sadly, he died in 2017.

My thesis advisor, Charlie Goebel was a hermit genius, very shy, but tough and scientifically libertarian (sink or swim) as an advisor. He received his BS degree from the University of Chicago at eighteen. He stayed on to obtain his PhD mentored by, Gregor Wentzel, now my academic grandpa. Wentzel had been a PhD student of Arnold Sommerfeld, known mostly for his books and his many influential students. Sommerfeld had sired (academically) four Nobel Laureates (Werner Heisenberg, Peter Debye, Wolfgang Pauli, and Hans Bethe). Goebel went on to be a postdoc at Berkeley, then become a professor at Rochester before coming to Madison. He was only four or five years older than I, but a full professor. He was already bald with a fringe of red hair circling the sides and back of his head.

I would screw up the nerve to venture into his office and ask him a question about quantum field theory, the subject of my research. He would spin his chair around, his back to me and face to the window, thinking, I guess. Eventually, I would fumble the question out and he would answer in great detail. I would thank him and leave. A few days later I would return to his office and announce that I had figured out the answer to the problem. I would produce my solution on the chalkboard in his office. He would say, "Yeah, that's what I told you." I would then slink out. I took graduate quantum mechanics and quantum field theory courses from him (the latter I sat through a second time). His lectures were often incomprehensible, but elegant in their way. They were always comprehensive and original, never from a textbook, but often from the live literature. I still have the notes. Sometimes he would say, "Forget everything I said yesterday; here is the correct way to do this or that calculation." Later he said he usually prepared them the day or night before, working "hand to mouth."

In the summer of 1966, I decided that I had done enough work to finish up. I learned in the spring that Charlie would be away for the summer in Boulder at a summer program, the same program I would attend a year later, myself as a postdoc at Penn. Yes, I remember flying out to Boulder accompanied by Jane and with two-year-old Michael asleep on my lap. He peed all over my lower half as the plane came into Denver.

I told Charlie that I thought I had done enough, and that I would quit working in May and begin writing it up. He asked, "Oh, what have you done?" I choked out what I thought I had done, and he said OK. I held up my sigh of relief until I left his office, red-faced. Again the academic libertarian thought such decisions were up to the candidate. You could sink or swim with your own decisions. Your reputation was yours to make. My dissertation was not the best, but it was all mine. I had received essentially no coaching. What I really learned was how to do things by myself. I did not have it down pat just yet, but it was a good step in the right direction. I had learned to fall on my face occasionally, pick myself up, and try again, until I got it right. Throw enough crap at the wall, and something is bound to stick!

Even though I was scared of him and his towering intellect, I cherished his being my mentor, because he did not lead me by the nose through my dissertation research. It is more and more common in these current times of research grants ("contracts" are even worse) and accountability to the grant monitor (or a tenure and promotion committee) to lead the student, step by step, through the project, leaving the student no opportunity to fail a few times and thus learn how the science process really works. That process always leads to a publication for the professor, but it is not necessarily good for the student's development.

I finally finished writing my dissertation in time to take the job at Penn by that September. First, there was the dissertation defense, an oral exam before the student's committee. I had asked Saul Epstein to be on my committee, because I had heard that he was interested in one of the techniques in the paper (perturbation theory). Epstein was a formidable character. Like Hal Lewis, he had been a student or postdoc of J. Robert Oppenheimer, earlier, the leader of the Manhattan Project at Los Alamos, New Mexico. Oppenheimer had assembled a galaxy of the world's greatest theoreticians in Los Alamos. Along with many refugees from fascism in Germany, Italy, and especially Hungary, they created the atomic bomb. "Oppie" went on to become the director of the Institute for Advanced Studies at Princeton (IAS). No dumbos or slouches worked for Oppie, at least not for long.

My graduate committee members received a copy of my dissertation a few weeks in advance (it was typed on an old-fashioned typewriter from my hand-scribbled text, with zillions of equations and graphical symbols). When I walked in, Epstein said, "This is about the most poorly written dissertation I have ever seen." The room was dead silent, and I felt twelve inches high. But we proceeded and much to my surprise (probably out of pity), Charlie came to my defense in clarifying a couple of things, which I understood, but left me tongue-tied. Finally, it was over, and I was sent down the hall, where I stood nervously reading posters on the bulletin board as they were probably comparing me to a turnip. The hallway pose of the despondent candidate is one I have seen at least a hundred times as a professor myself. Suddenly, out pops Epstein, congratulating me. I thought he was going to ask me to completely rewrite (or hire a ghostwriter for) the dissertation, but there was no such demand. I had suffered the appropriate amount of humiliation and test for resilience—I was free at last.

Perhaps Charlie sympathized because he was a terrible writer himself. He did not write many papers over his career, although a few were highly cited. He was coauthor of one titled "Topology of Higgs Fields" published in 1975 that received 331 citations. This paper was published in *The Journal of Mathematical Physics* decades before the actual spectacular experimental detection of the Higgs Boson. Charlie communicated more by letters to other authors than through conventional scientific papers. He liked to read books and took extensive notes on yellow legal pads with an old-fashioned fountain pen. Each was stapled and stuffed in a folder that went into a file cabinet. He was a critic of many papers. The last time I saw him in 2007, he was retired, and he was complaining about the weather person who writes for the local paper in Madison. He asked me (as a now-atmospheric scientist) about the thick layers of ice found on clotheslines during winter: "Was this layer caused by super-cooled water drops hitting the subfreezing wire?" I said, "Yes, you are right," and he replied, "Good. I already wrote to that guy and told him he was wrong!" Charlie might be approaching a crank phase. Maybe I am, too.

My years at Madison were exciting with hard studying, but with many parallel pleasures as well. I can only do theoretical physics for so many hours per day. Many of my experimentalist friends spent late nights in the lab. Instead, Jane and I watched many movies on weekends; Ingmar Bergman was in his prime as a director, and there was a theater in town that featured that kind of film. We became Wisconsin Badger as well as Green Bay Packer fans and watched the games (Badgers live at home or on TV on the road, Packers on TV) during the fall seasons. On Fridays or Saturdays, we often played bridge with other graduate student couples. We were pretty mediocre at the game, but the beer was good.

My oldest son, Michael, was born May 25, 1965, in the University Hospital. Jane received her MS in psychology and continued to work as a research assistant until Michael came along. We took road trips out West for vacations a couple of times. The first trip was in our 1958 two-door Chevrolet. I removed the back seat, and we stuffed the floor with magazines and newspapers for padding (leveling out the hump due to the drive train beneath) and we slept on that in campgrounds in Yellowstone and Grand Tetons National Parks in Wyoming and the Rocky Mountain National Park in Colorado. The second time to Glacier National Park, we sprang for a tent.

## Postdoc in Philadelphia

I spent two years as a postdoc at Penn. This was in the David Rittenhouse Laboratory on Walnut Street near 30th Street Station, and a few tens of blocks from historic downtown Philadelphia. In our first year, we lived in a row house (elsewhere it might be called a townhouse) in Upper Darby. It was an easy streetcar ride down Market Street to my office. Jane had a part-time job nearby in the Psychology Department working on a research project for a professor there. I wanted to know what it was like living in a row house. They form very picturesque scenes in the East Coast cities of Baltimore and Philadelphia. Sadly, we had a spouse-abusing drunk next door. As soon as our twelve-month lease expired, and we had the assurance of the second year of my appointment, we moved to a duplex in Lower Marion on the mainline. This is a famous train line with a station about eight blocks from home that delivered me to 30th Street Station, a few blocks from my office. Lower Marion had one relatively modest neighborhood where we rented one side of a duplex. The mainline connected the dots of the richest suburbs of Philadelphia. My morning train ride was shared with people reading their *Wall Street Journal*. I wore a sport coat and tie every day and carried a neat black leather briefcase.

Our duplex neighbors were avid roller derby fans. It had been a while since I was friendly with folks from the same stratum of society where I originated. In case you have never heard of roller derby, it is a professional sport in which there are different leagues for men and women. It is a bit like professional wrestling, only on skates. There is a circular track, and the team members go around the track at high speed, pushing and bumping their opponents. One player is designated the jammer, who tries to lap the other skaters, while his or her teammates are helping in just about any way they can. We went to a couple of these spectacles in which the crowd, fully beered and pretzeled, has a loud old time cheering its favorite skaters. Many short games are played over the evening with female and male teams alternating. It is a wildly popular sport in Philly. I have scarcely heard of that sport since.

Actually, we leaned toward the cultural side of entertainment offered in Philly, trying to become East Coast sophisticates. Several professors and I walked to lunch at local sandwich shops. The conversation focused on the latest art exhibits, plays, and musical performances. Naturally, I sought to fit in—a struggle considering my hillbilly origins, although I did take a music appreciation course at UTK that at least gave me sufficient vocabulary to survive. I recall that we bought season tickets to the Philadelphia Symphony. Sadly, the concerts were on weeknights, and I was often so tired from my day at the office that I tended to fall asleep. Those two years constituted my high point for cultural participation as an audience member. Frankly, I prefer a good baseball game.

I had wonderful senior mentors such as Ralph Amado, Sidney Bludmann, Henry Primakoff, and Abraham Klein. Unfortunately, just as I was moving to Penn, two of the people who had common interests with me had already skipped away to MIT. Amado and Klein had made significant contributions in my area (quantum field theory of mesons, especially analytical solution stuff), but had dropped it to work on theoretical nuclear physics (I took it that they had peaked and moved to something more easily funded and less competitive). Of the two who left Penn for MIT, one died tragically and the other did not gain tenure at MIT. There were two assistant professors at Penn who were somewhat interested in my area, one coauthored a paper with me, but both were denied tenure at Penn after I left.

This flushing out of physics faculty at such strong institutions was partly because of the sudden war-induced drought of federal grant money, and many assistant professors did not receive their basic nine-month salaries from the budget of the institution. Part of their nine-month salary was leveraged by outside grants from agencies such as the Department of Energy. As the Vietnam War proceeded, President Johnson decided to pay for the war by shrinking the budgets for some of the departments instead of hiding it as was done in the more recent Iraq War. This is an important lesson for departments of atmospheric sciences who have often leveraged tenured or tenure-track faculty salaries with grant money in times of plenty. What goes up can go down.

My two years of study at Penn were at times overwhelming. My office mate, Julian Noble, had been an undergrad at CalTech and a graduate student at Princeton, both far above my grade. He was brilliant, funny, sociable, and knowledgeable. Unfortunately, he was a terrible office mate because he had no understanding or respect for ordinary mortals like myself. He also lived on the mainline and took the train in. He often slept late and ran to the train station, shaving his heavy growth of whiskers in the office with his electric razor. He came in every morning with a couple of opening jokes. He told me that he had read somewhere that this was a good idea, and he read a few jokes in advance as preparation every day. He tried them out on me before on our betters.

Julian often entered the office, looked over my shoulder to see what I was working on, or merely looked to see what I was reading. This usually started an endless lecture on that or whatever subject entered his mind as he expounded. He was hyper. Israel humiliated the Egyptian Army in the 1967 war. It was as if the war was occurring just outside our office. Of course, I was sympathetic with the Israelis, but Julian could not stop his macho praise for them to me and on the phone with whomever he could find. In short, Julian was really hard to take and at the same time get any serious work done. I loved the guy, but I could not take it much longer.

I asked to be moved in the second year. My elders knew Julian well and sympathized with me. My office mate became Dr. Mohinder Khanna, a very pleasant Indian theorist whom I really enjoyed and admired. He went on to be a professor at Punjab University at Chandigarh in India. Julian obtained a tenure-track job at Penn, and he deserved it because he was really a brilliant and productive physicist. He was denied tenure at Penn and spent the rest of his career at the University of Virginia until his death in 2007 at the age of sixty-six. I regret that I never communicated with Julian after I left Penn.

During my second year in Philadelphia, our daughter Cheryl was born at the University of Pennsylvania Hospital. We were lucky with hospitals for the kids, both of the university hospitals were great, and the expenses were minimal. Michael was well, but the winters in Madison and Philadelphia were hard on his ears, which led to infections that accompanied the common respiratory illnesses of Pennsylvania's wet, gray winters. I had enough flexibility at work that Jane and I could alternate or go together for the doctor visits. I was told that I had many of the same disorders as a baby. In fact, ear infections have taken their toll on me. When we first arrived in Madison in 1961, I had a staph infection in the right ear that dragged on for more than a year. I suffered another infection in 1968. A huge shot of penicillin left me allergic to that drug. I still have a gaping hole in the eardrum on the right side. Now I wear hearing aids that amplify across the spectrum on the right and emphasize the high frequencies on the left.

The job market in physics was in a death spiral. There were too many PhDs out there looking for teaching jobs, and the great boom of building red-brick branch campuses of the great state universities with job vacancies was coming to a close. The states were also cutting back because of the war. In the spring of 1968, I started sending out CVs to anyone I could think of, not only to universities and colleges but even to labs like Oak Ridge, where I learned that Herman Postma was director of ORNL. Postma was a fresh PhD from Harvard when I had worked in the Thermonuclear Division, and I had known him then. He had risen rapidly through the ranks in a very short time. And of course, there was no call for a theoretician in my game, even if I wished to go back to plasma physics, which I could easily have done. Actually, I think ORNL had lost its stake in that field to Livermore National Laboratory, Princeton, and elsewhere. Of my old friends and mentors, Ken Fowler had removed to Lawrence Livermore National Laboratory (LLNL), and Bob Mackin had scampered to JPL. Incidentally, Fowler became head of the Nuclear Engineering Department at Berkeley and a member of the NAS, while Mackin became part of the senior leadership team at JPL.

I had no other offers, and I thought I might have to apply in industry, perhaps even in the defense sector. I was not keen on the last choice, but I am a patriotic person, and I would have served even in the active military had I been called. I never was called throughout the Vietnam era. I was rated 3-A in the draft, possibly because I was married and a little older. Many of my friends in graduate school agonized over the lottery, but none that I knew of were drafted.

## Physics Professor

Next came a great break for me that had everlasting effects for my destiny. My best friend from graduate school, Phil James, had just landed a job at the University of Missouri-St. Louis (UMSL). It was getting late but at the last minute, there appeared another opening on that campus.

He suggested me, and I boarded a plane to interview. When I arrived, the entire campus consisted of one building, Benton Hall, plus the old clubhouse—the campus in the northwest suburbs of St. Louis was built on a former golf course, the Bellerive Country Club. The physics, chemistry, and psychology departments are still housed in Benton Hall. The enrollment was a few thousand, and during my time there, it rose to about 11,000. In 2013 it was nearly 17,000. As I taught physics over the decade there, new buildings sprouted all over the former golf course and around the pond. The horrible job market allowed us to hire way over our heads. One of our first was Ta-Pei Cheng, a theorist in my old field of high-energy physics, who had a PhD from Rockefeller University in New York City under Abraham Pais, then a postdoc at the Institute for Advanced Studies for two years, then back to Rockefeller for two more years. Ta-Pei continued to work as a high-energy theorist and eventually became emeritus at UMSL. Ta-Pei was a brilliant scientist and a gentleman. His mentor, Pais, was a great theoretician, but he is even more recognized as a historian for his excellent biographies of Einstein, Bohr, and others as well as memoirs of his own interesting life (Pais 1997).

I had a total of ten enjoyable years connected to UMSL. Aside from my old friend Phil James, an extremely important friend and colleague was Jacob "Jake" Leventhal. Jake and I were very close, having lunch together nearly every day. We plotted how to make the department better. We worked to hire good people and promoted St. Louis as a good place to live—and it was.

While at UMSL, I wrote papers with both Phil and Jake. I was beginning to see the end of the line as a high-energy physics theoretician. It could not work for me in the small college-like atmosphere that we had—not competing with a workaholic, monkish postdocs at big schools, who did not teach, and some of them into their fifth year as postdoc researchers and desperately competitive. I thought of theoretical chemical-physics the theoretical counterpart to Jake's experimental research—and we did publish a few papers together in that area. The year I was accepted to take the sabbatical at NCAR, I also had the same opportunity to go to Brookhaven National Laboratory (BNL) in the chemical-physics role. I suspect that if I had gone that way, I would have wound up running huge computer programs, and I was not so keen on that given my weakness as a programmer. The scientific area I would have entered at BNL has turned out to be important in theoretical chemistry. Nevertheless, Jake and Phil made me a better scientist and person over those years.

On August 24, 1970, at about 3:45 a.m. one morning, there was an explosion on the campus at Madison just outside Sterling Hall, where I had worked. The Vietnam War became the subject of demonstrations across the country, especially on college campuses. This one was a car bomb loaded with explosives, set up just outside of a building on campus. It was intended for the Army Mathematics Center that shared some space in the physics building, Sterling Hall. The Center was not harmed but several graduate students were working in the solid-state laboratory at that odd hour in the night. One, Robert Fassnacht, was killed and three others were injured. I knew Robert as a young graduate student there as I was nearly finished. Those were tough times, no doubt partly because of the draft. Many American young men left the country for Canada to avoid it. My own support for the war was flagging by that time, and I could not see how it would be productive to go on with it. Recent revelations about Nixon's nefarious behavior in manipulating the war negotiations for his own political purposes supports my changing position back then. Most say we lost that war, but I think it might have been a moral victory since communism did not follow the domino effect after we (ingloriously) departed. Of course, the price we paid in flesh, treasure, and national morale was extremely high.

### Politics

Over the years, my sympathy for the "little guy" grew. I recall in my university days, so many of my childhood friends had a hard time trying to get out of the lower-middle or even lower class. Such ridiculous requirements as ROTC and physical education (mostly providing jobs for athletic teaching assistants, whom I resented) placed an undue burden on them. Those folks had the ability, but inadequate high school preparation and cultural barriers prevailed. After taking a few economics courses to satisfy social science requirements, my sympathies amplified. Despite these feelings, I was and am patriotic, and I was slow to oppose the Vietnam War.

I strongly supported the losing cause of George McGovern in 1972 against Nixon, whom I already disliked intensely. I attended local Democratic Party meetings, and I was McGovern's precinct captain in the cause.

I knocked on all the doors in my middle-class neighborhood of University City, a close-in pluralistic suburb to St. Louis. I actually saw McGovern in a rally in St. Louis. Of course, he went down in a landslide. It was the summer of the Watergate scandal that eventually led to the resignation of President Nixon. He was sure to have won the election without any of the crimes of Watergate or even his unnecessary deception in the war that he had used to enhance his chances of overwhelming McGovern. My experience with McGovern taught me to avoid supporting similar candidates like Bernie Sanders even though I might lean toward their views. Politically, I was moving toward a liberal pragmatism.

Later there came the Star Wars project that had proposed implementation of a mechanism to destroy incoming nuclear-armed missiles from our Soviet adversaries. Many of us in the physics department prepared and gave lectures to show how much damage would be caused by a nuclear encounter with the Soviets and in addition how futile it was to pursue the Star Wars project, which seemed to me to be a misguided attempt to do what was physically impossible and very expensive to boot. The United States continued working on the scheme, even though most physicists in the West thought it would never work. I understand that Soviet physicists agreed. But leaders on both sides (Reagan and Brezhnev) thought it might work. In the end, the Soviets stumbled into a hopeless and expensive war in Afghanistan. Already stagnating, the Soviets essentially went broke and were on their way to collapse and dissolution. So I guess Reagan was right, and I was wrong.

My colleagues and I were active in university politics, always trying to make UMSL a real university—really, an impossible task for historical, political, and geographical reasons. The school and our department were so young that even assistant professors had a strong influence on standards in hiring and curriculum. The school was unique in that it had no graduate students (at least then) and we could tailor things for ourselves. There was no engineering, so we made a pre-engineering track for those who wanted to transfer to the bigger campuses in Columbia and Rolla. We created an undergrad applied physics program with a senior year that featured applied courses, such as fluid dynamics, statistical physics, and finally, for my sake, atmospheric physics, which I was beginning to learn. We also had special topics courses for our regular physics majors, such as relativity, kinetic theory of gases, and many others. We soon had a very good faculty, and with no graduate students, we tried to develop courses that would be challenging to teach and interesting to take. I have always liked to teach subjects I knew little about. That was an incentive to learn that subject. You can easily see how I began to learn about the atmosphere, ocean, climate, the space program, thermodynamics, and mathematical statistics. Then there are my hobby subjects: Russian history and literature, history of science, history of philosophy, English and American history and literature, biographies, and more.

The department needed more enrollment, and we capitalized on the creation of new courses such as astronomy, and I cooked up a course on introductory meteorology. I repeat the phrase from my old advisor, Charlie, "teaching from hand to mouth," referring to learning the stuff the night before having to teach it. I am an obsessive autodidact.

Jake had his undergrad degree from Washington University in St. Louis, a PhD in physics from Florida, and a postdoc from BNL, far out on Long Island. He was an excellent experimentalist in chemical and molecular physics. He came to UMSL a year before I did. He had built a well-furnished research laboratory in which he could create beams of molecular ions (one or more electrons stripped off) and aim them at a volume of target molecules to see how various species interacted through scattering. I became interested in this work, and we wrote several papers together in that field. I could never quite understand all the intricate details of the laboratory, but we were a good team, connecting the lab results with my knowledge of quantum mechanics.

Although Phil and I wrote a few papers in theoretical high-energy physics, we both thought it might be a good idea to move away from our old dissertation field and try something new that would be more fulfilling. I especially felt the need to make a change. As much fun as it was teaching physics to undergrads, there had to be a way to move to a research discipline that I would find influential and the work rewarding. I wrote an application to NCAR, as discussed in chapter one. I was declined, but since my second try was for a sabbatical year in which half of my salary would be paid by UMSL, I received the offer to spend twelve months at NCAR in Boulder. As a fallback, I applied at Brookhaven to work in chemical physics related to the theoretical side of the field in which Jake worked. I turned down the Brookhaven offer because I was concerned that most of the work being done by theorists in that field involved heavy use of computers. I took the chance that I could avoid exposure of that weakness in the atmospheric sciences. I had not yet discovered that I wanted to specialize in climate science.

This decision to go to Boulder instead of Long Island was hard for Jane. I am not sure she ever forgave me. She looked forward to being close to New York City for a year. She became interested in politics at the time and became a campaign manager for a candidate for St. Louis County Council. The candidate happened to be our next-door neighbor, Betty Van Umm, who was a young widow. Betty was a gifted politician, and we agreed that we could rent our house out, and Jane could room with Betty, who happened to have kids that were the same ages as ours. Betty won her race, and after serving one term she became a senior officer for public affairs and economic development at UMSL.

During the summer of 1974, I stayed home with the kids while Jane worked on Betty's election. I went to Boulder and took both kids for the fall semester of the school year, one for the spring semester, and the other for the upcoming summer. After the year in Boulder, I returned to St. Louis to take up teaching again. The sabbatical leave and its complications led to our separation. I found that I did not like politicians, at least not at close range. The intensity and passions of campaigns and the post-victory celebrations were too much for me. Moreover, it was the peak period of the rise of feminism. While completely in favor of its tenets, I was subjected to social pressures that I was not able to enjoy. Jane and I sought help from a marriage counselor. The result was that we separated, and I rented an apartment. I had reasonable visitation times with the children, but at their age, my visits with them were problematic. The visits drifted into meals at fast-food joints. It is painful to remember the experience.

In December 1975, I received notice that my two papers had won the NCAR Outstanding Publication Prize and a check for $500 was enclosed in the letter. This was broadcast throughout the meteorology community, and I was immediately known for this recognition. There were about one hundred journeymen PhD research scientists at NCAR, all of whom were writing papers. Somehow I was nominated, and I won. I could not attend the ceremony at NCAR because of teaching, and I asked Phil Thompson to accept the award on my behalf. Later, I was to give a lecture on my work in the big auditorium at NCAR. Winning the prize now opened the door to many opportunities. Soon I was being invited to universities to present seminars, and I was giving invited talks at conventions.

Jerry (left), Michelle Fieux (right), and St. Louis colleague Phil James (middle) at an outdoor restaurant near Nice, France. The occasion was a meeting in Nice, ca. 1979.

Jerry (left) and St. Louis colleague Jake Leventhal (right) in St. Louis, ca. 1978.

Woods Hold Oceanic Institution Summer Program 1976. Group photo of the lecturers and students in front of Walsh Cottage. With Permission from WHOI.

Group shot of attendees at a conference at Lamont-Doherty Geological Observatory on Paleoclimatology. Well-known scientists Manabe and Imbrie can be seen in the front row, ca. 1978.

# CHAPTER 4

# WOODS HOLE AND PALEOCLIMATOLOGY

In this chapter, I take the reader to the southern coast of Cape Cod in Massachusetts. There are two large scientific institutions located in the small town of Woods Hole. One is the Marine Biological Center, which I will not discuss, and the second is the Woods Hole Oceanographic Institution (WHOI), which is of great interest for climate science.

## Woods Hole

Woods Hole Oceanographic Institution (WHOI) is located on the southern coast of Cape Cod, Massachusetts. It inhabits a small campus and owns some research vessels at the pier. WHOI collaborates with the Massachusetts Institution of Technology (MIT), and a number of its senior scientists are MIT professors, who teach graduate courses and mentor students on the WHOI campus. WHOI gets most of its funding from NSF, but some from other agencies interested in oceanographic research for applications. The research mission of WHOI spans the entire range of oceanographic problems, from the ocean floor to its surface, ranging from the distant past to the future, on topics from chemical composition and processes of the oceans to the dynamics of flows, all in both theoretical and observational modes. The ocean plays a huge role in climate science because it covers 70 percent of the planetary surface, it is a reservoir for all kinds of chemicals relevant to climate science, and the details of ocean circulation and the general architecture of its influence lags in the climate ranging from months to millennia (see the WHOI website for more discussion). Then there is sea level.

In the spring of 1976, I received an invitation to spend the summer at WHOI. The occasion was the Geophysical Fluid Dynamics Program, a summer lecture and research program that sponsored about twenty graduate students and about thirteen faculty lecturers/mentors. I was asked to be one of the faculty advisors. The theme of the program was *Global Climatology*. The organizer of the program that summer was Prof. Andrew "Andy" Ingersoll, who was (and still is) a planetary scientist and professor at CalTech. Andy and his family kindly greeted me and invited me over for dinner on the first day. We had useful and interesting conversations that evening and throughout the summer. I did not see Andy again until about twenty years later when I was visiting CalTech for a meeting on another topic. Stephen Leroy, a former student of Andy's and for many years a research scientist at Harvard, had become a collaborator of mine on a satellite experiment that was to make use of the Global Positioning Satellites (GPS). At the time, Stephen was working at JPL. He took me over to an informal seminar being given by a CalTech graduate student. I sat down beside Andy, and he did not notice or recognize me. I was wearing a badge because of the workshop I was attending elsewhere on the campus. I scooted the badge in front of him, and he jumped in his seat and put his arms around me. After the seminar, we retired to his office where on the wall was a group photo of the WHOI summer program from the summer of 1976 with both of us and all the other participants.

In addition, there were twelve visiting lecturers, mostly from the New England universities who dropped in for a few days. (Please excuse my constantly listing notable people, but many of the readership will recognize them.) These included Wallace Broecker, geochemist and paleoclimate scientist from Lamont-Doherty Geological Observatory, Columbia University; Peter Gierasch, planetary atmospheres, Cornell University; John Imbrie, professor, marine geologist, and paleoclimatologist, Brown University; Richard Lindzen, geophysical fluid dynamics theoretician and climate scientist, Harvard, later to MIT; Abraham Oort, atmospheric scientist from the Geophysical Fluid Dynamics Laboratory (GFDL), Princeton; Barry Saltzmann, professor, atmospheric and climate scientist, Yale University; Tom Vonder Haar, professor, satellite retrieval of the climate system, Colorado State University; and Pierre Welander, oceanographer, University of Washington. I got to know all of these people.

The students were each to do a project, write it up, and give a talk at the end of the program based upon their work. I was assigned four students: David Pollard, a grad student of Ingersoll's at CalTech, later a distinguished senior research scientist in paleoclimatology at Penn State;

Bruce Wielicki, at that time a grad student at Scripps, later a highly decorated senior scientist at NASA/Langley Research Center; Glenn White, soon to be a grad student of Mike Wallace at University of Washington, and just retiring after thirty-four years at NOAA; and Lawrence Kells, a grad student of Steve Orszag in applied math at MIT. I presume that Andy Ingersoll made the assignments. I drop these names since I was flabbergasted at the amount of talent and responsibility handed to me. The program and summaries of the lectures based on notes taken by the students are still available online (darchive.mblwhoilibrary.org/handle/1912/1537).

What a cast of stars to hang out with for eight weeks in the summer of 1976! Woods Hole was another of those magical places in the summer, although in the last week or so we were concerned by a hurricane creeping up the East Coast, and everyone on the Cape had to prepare for it. The storm missed us, moving out into the Atlantic as it passed, but there was some wind and rain.

Strolling with a friend or two in the village of Woods Hole at lunchtime, breathing the sea air, looking at the boats, and sniffing the smells of seafood, wafting from the touristy restaurants along the way added to the exhilaration. In summer the town is a tourist haunt with the usual souvenir shops and related businesses. There were street vendors of hot dogs, pretzels, and other finger foods throughout the day.

WHOI staff rented a room for me in a boarding house a few miles away in the small town of Falmouth. I had driven my 1965 Dodge Dart from St. Louis. The house backed up to a pond that was perhaps a half-mile across. Such ponds are common on Cape Cod, reminding us of their creation by the Laurentide ice sheet in the last ice age. Sitting in the yard on the slope above the pond made for relaxing evenings before sundown. By dinnertime, I was usually bone-tired from absorbing and processing information as well as supervising the graduate students. I had arranged for Michael, then age eleven, and Cheryl, age nine, to join me from St. Louis separately for a week each. Cheryl came during my last week and rode back to Missouri with me in the car. I took one on a Sunday ferry trip to the island of Martha's Vineyard and the other to Nantucket Island, each an hour or two away on the water.

While I got to know many of the staff and visiting scientists in the summer program at WHOI, I was particularly interested in learning more about paleoclimatology and its mighty players who were present, especially James D. Hays, John Imbrie, Sir Nicholas John Shackleton, and Wallace Broecker, all of whom were very kind and friendly toward me. The field had been growing in the last few decades and finally came of age in the spectacular year of 1976. It was the very year of the groundbreaking publication of Hays, Imbrie, and Shackleton in the journal *Science*. This was one of the most important papers in the geosciences in the twentieth century. But before we get to that, let's have a look at what this field is all about, how the research is conducted, and finally, its importance in the broader field of climate science. I was so lucky (once again) to be around during many intense discussions of these topics. The lectures were exciting with fascinating questions and occasional heated exchanges. It must have been a supercharger for the students.

## Paleoclimatology

*Paleoclimatology* is the study of ancient climates on Earth. We have lots of evidence of climate change over the 4.7-billion-year history of our planet. For starters, we now know some things about the geographical configuration of the continental landmasses—the continents have combined, separated, and moved about endlessly on time scales of tens of millions of years. There are reasonable estimates of the composition of the atmosphere and oceans for the early planet. The information flow (fossils!) steps up when life appears. There is a mega-burst of living species that occurred about 500 million years before present (beginning of the Cambrian Epoch). Living matter turns to fossils, and these preserved markers provide all kinds of clues as to conditions at the time of their deposition. The study of these ancient life forms is called *paleontology*. Almost all paleontologists were trained as geologists or biologists, with a few archeologists.

Today's paleoclimatologists come from a broad range of home disciplines, often sorted by their specialized training in various techniques of inferring past climates from observational evidence. For example, many are marine geologists looking at the sea and the seabed. Glaciologists can be trained in geology, hydrology, engineering, geography, or even physics, and many glaciologists consider themselves paleoclimatologists. The ascent of paleoclimatology was ahead of the growth of all of climate science. It is a critical part of climate science because it provides context to our understanding of how climate has changed in the past. In order to know what future climates might be like and even where our present climate came from, we need to know what it was like in the past. Moreover, evidence from the past can play a role in testing our theories and models. To predict the future and fully understand the past, our model simulations have to be consistent with the evidence from the past. Testing is actually not realistic because of the many unknowns, but fitting the discoveries into the general web of our understanding is crucial.

Aside from my curiosity about the subject, my interest in paleoclimatology stems from the possibility of the applicability of my class of small climate models to simulating past climates and understanding why they changed from one state to another. What were the forces that caused the changes, and could models help to figure this out? The huge expense of using general circulation models (GCMs) prohibited them from playing this game in the 1970s. As GCMs became more sophisticated through continued improvements in computers, there was a place for the simpler models because they could be easily understood, and they held the promise of easy insight into paleoclimate processes. Moreover, they could be used as pilot studies for future experiments with GCMs. These reasons were the basis for inviting me to WHOI in 1976.

## Paleoclimate Data

Data used in studying paleoclimatology depend on the time span of interest and the resolution in that span. In the case of the ice ages of the last few million years, we might consider time spans of a few tens of thousands of years and a time resolution of hundreds of years. Paleoclimate data is usually in the form of layers that were deposited time after time. Examples include pollen layers from ancient lake beds, ice layers from ice sheets where the annual cycle is identifiable, tree rings where the outermost ring is the most recent annual layer of wood growth; coral layers are similar to tree rings in that outer layers are the youngest. We need a means of telling when the information was laid down and what evidence there is in the particular layer that might indicate conditions at the time.

## Cores from the Sea Floor

The data of interest at WHOI and Lamont-Daugherty Geological Observatory is mostly from deep-sea cores. In this case, a ship with deep-sea drilling capability is sent out, and cores are dug from the seafloor. The youngest material in the core is at the top, older layers are found the deeper you go.

Ocean drilling started in the 1940s but got underway seriously in 1961 when "dynamic positioning" made it possible for the ship to remain steady in fast or shifting currents. In 1966 the drilling ship *Glomar Challenger* began coring for scientific purposes. In 1983 the drilling program moved its headquarters to Texas A&M (just three years before I arrived there) and the main drilling ship was the *JOIDES Resolution*. The effort during this period became truly international, with twenty-six participating nations. I have enjoyed taking my students on tours of the International Ocean Discovery Program (IODP) facility on the east end of the Texas A&M campus. I especially enjoyed befriending and working with the inaugural and long-time IODP director, Phil Rabinowitz, both in this role and later as we were fellow department heads and jolly warriors for lost causes in the College of Geosciences.

An experiment is conducted as follows. First, a science team goes to sea on one of the program's drilling ships. The *JOIDES Resolution* is the US ship used in many cruises. Most of the cores recovered since 1982 are stored in an archive at Texas A&M. An individual core may be hundreds of meters long. When a core is examined, a segment is selected from it that represents a period of time in the past. This segment is then cut along its length to form a half-cylinder. The half-cylinder is loaded into a tray so that a chronology can be established. One can see different shadings and colors reflecting chemical changes along the core. These features can be accurately measured and chronology can be established. Along the core, going back in time, there are sediments containing many different kinds of indicators (e.g., thin layers of ash from volcanic eruptions, wind-blown dust, magnetic polarization of particles that have fallen slowly down the water column, etc.). Oxygen isotopes are especially interesting, but isotopes from other elements can also be used as indicators of ocean properties such as temperature and total ice volume on the planet.

These indicators or *proxies* give us information about the past, but it took many years of research to pin down exactly what they were telling us. Skeletons of microscopic critters (species) embedded in the sedimentary layer form a record similar to the pollen record from a lakebed. Similar to pollen studies, the abundance of different species (here foraminifera [skeletal material usually made of carbonates], radiolarian [skeletal material usually made of silica], etc.) can be used to indicate various sea properties such as ancient temperatures and salinities. Isotope studies of the tiny skeletons with mass spectrometers (called *tests*) can be used to infer not just properties of the whole water column but that some species are top dwellers and some are bottom dwellers. In the case of top dwellers, larger animals eat these tiny organisms up the food chain, and the feces from these larger biota sink to the bottom where they accumulate. Differences among these species can be used to discriminate ancient sea surface temperatures from deep-water (*benthic*) properties based on the type of organism that is being examined. I will return to some of these studies shortly. An oxygen isotope concentration can tell us about the seawater that was taken into the building of a micro skeleton, for example.

Much planning is necessary before an expedition is approved and implemented. Some studies seek locations where the sedimentation rates are high. This helps resolve the chronology, providing more data per unit of time in the past. Cores from drilling the ocean floor can be used for many other purposes, to mention just one, finding the direction of the Earth's magnetic field at different times in the past. Magnetic field reversals revealed the spreading of the seafloor and ultimately the evolution of the revolutionary theory of continental drift—the theory was clinched by ocean drilling data and analysis led by scientists at Lamont-Doherty Earth Observatory in the 1960s (see ldeo.columbia.edu/about-ldeo/history-lamont).

Broadly similar methods are employed in examining ice cores that can be drilled from the surface of ice sheets. In that case, each annual layer of snow that later turns to ice can be identified, and these layers can be used to measure time in years going down the ice core. The composition of the layer holds all kinds of information such as whether or not there was a volcanic dust layer that year, readily identified by the sulfuric chemicals. Often the isotopic ratios of certain atoms such as oxygen or hydrogen give clues to the temperature at the time of the deposition of the snow.

Ice cores are especially useful for studies spanning only a few thousand years (google.com/search?client=safari&rls=en&q=ice+core+samples&ie =UTF-8&oe=UTF-8).

## Paleoclimate Findings

Long and painstaking work beginning in the nineteenth century and conducted in the field by geologists and geographers indicated that large portions of present Great Britain were at one time covered with ice, along with evidence that the same was found to have occurred over much of Scandinavia. The North American ice sheet (called the Laurentide Ice Sheet) was at maximum surface area and ice volume twenty thousand years ago. It had terminated south of the present Great Lakes, covering even Cape Cod. Chronologies are established in various ways, including the identification of plants from fossil pollen data combined with radiocarbon and other kinds of dating.

Data suggested that the large-scale glaciation process seemed to repeat itself many times over long intervals of slow (many millennia) growth and rapid (a few millennia) decay. One could obtain rough estimates of the ice volume over time as glacial changes occurred. The waxing and waning of the great ice sheets were enough to raise and lower sea level because of the appreciable removal of water from the oceans and its subsequent deposition on the land as ice. For instance, sea level was 120 meters lower twenty thousand years ago, according to many different independent kinds of evidence.

It took decades but paleoclimatologists established that the great ice sheets grew to their peak volume about every one hundred thousand years. The growth phase was a kind of stutter-step of gradual ice buildup, and the retreat or decay phase was more abrupt, with wastage of the ice occurring only over a period of ten thousand to twenty thousand years or so. The cycles were not the same each time, but an overall pattern began to emerge. Growth was slow, decay relatively fast—evidence of nonlinearity in the process.

## Earth's Orbital Changes

Long before we had the sea core data giving precise chronological and ice volume data on the massive glaciations and their exact timing, theories had begun to emerge suggesting that there might be an astronomical explanation. It might be a physics problem. Changes in the orbital parameters include the *eccentricity* (how much does the annual trajectory deviate to an elliptical shape from a circle and does it change over geological time scales?), *obliquity* (or tilt angle, how much does the angle of the Earth's spin axis differ from a perpendicular to its orbital plane, and does it change?), and *perihelion* (the point or calendar day of the year on the elliptic trajectory of the Earth about the Sun when is it closest to the Sun, and does it change?).

There was considerable interest in the physics and applied mathematics communities over the nineteenth century on the problem of planetary orbits. If Earth were the only planet revolving around a spherical Sun, the orbit would be a pure ellipse as inferred from astronomical data by Johannes Kepler and later explained by Isaac Newton, both spectacular discoveries. Without external disturbance, the eccentricity, obliquity, and date of perihelion would be fixed. Because of the great mass of the Sun, the planets in the system go in nearly elliptical orbits. But they do perturb each other's orbital elements by tiny amounts because of their gravitational fields. The problem of solving these *many-body problems* is very difficult and purely analytical solutions are out.

The study of planetary orbits and their orbital elements lies in the field of *celestial mechanics*. In the nineteenth century, many mathematicians took up the challenge and learned how to find approximate solutions through the use of what is called *perturbation theory*. In this very elegant approach, one assumes the orbit is elliptical, and it is perturbed by a very weak influence, such as the orbit of a distant planet. These techniques and their approximate, hand-calculated solutions were used to pin down the orbits of the planets with incredible accuracy. To me, the level of accuracy in both the observations and the calculations indicates how incredibly well we could do these things a century ago.

To get an idea of how well the celestial mechanic guys and the planetary astronomers could do their jobs in the pre-supercomputer era, consider the orbit of Mercury, the nearest planet to the Sun. The observed precession of the perihelion of the planet Mercury was $574.10 \pm 0.65$ arc seconds per century or roughly one-sixth of a degree per century. The celestial mechanics at the beginning of the twentieth century sought to explain this precession using the perturbations of the trajectories of the other planets and the slight nonsphericity of the Sun and the planet, leading to a correction of 531.63 ± 0.69 seconds of arc per century. Taking into account the slight oblateness of the Sun's shape led only to less than one second of arc per century—still they were 43 seconds of arc per century short. This was something of a calamity for the Newtonian theory, which everyone at the time accepted as gospel. It was Einstein's General Theory of Relativity, published in 1915 that solved the problem with a correction due to the distortion of the gravitational field near the Sun. This theory led to the missing 43 seconds of arc necessary to shut down the clamor in the science community. Further confirmations of general relativity were to come shortly. I tell this story to indicate the importance of the theory and measurements, and to relate it to one of the key revolutions in physics at the time. In fact, this and later confirmations overthrew the conventional wisdom held for more than a century about the limits of Newtonian mechanics.

In the modern era, we can do much more with celestial mechanics because of the high-speed digital computer and the demand for accurate information related to the orbits of artificial satellites. Moreover, orbital mechanics is useful in the never-ending demand for defense-related applications (e.g., trajectories of intercontinental ballistic missiles and satellites).

The most interesting theory of the ice ages came from a Serbian, Milutin Milankovitch. He developed a theory linking celestial mechanics and the ice ages. WWI was breaking out, and Serbia was in it from the beginning. Milankovitch was arrested, but through personal connections, he was placed under a kind of house arrest in Budapest with the right to work in the library of the Hungarian Academy of Science. He continued his work on climate change and celestial mechanics.

Over a period of decades beginning with his house-arrest period in Budapest, he calculated (by hand) tables of the rate of solar energy received at different latitudes as a function of time of year for different values of eccentricity, obliquity, and the phase of perihelion. He found that the eccentricity had a dominant period of 100,000 years, the obliquity had a dominant period of 41,000 years, and the phase of perihelion had a split dominance of 23,500 and 19,000 years. His work did not gain much favor until many years later, and sadly, well after his death. I have contributed a chapter to a volume commemorating the anniversary of his death.

The most recent calculations relevant for this story involve the gravitational perturbations of Earth's orbital parameters due mainly to Jupiter. This work was carried out by André Berger and his students on modern supercomputers at Université Catholique de Louvain in Belgium. These state-of-the-art calculations were published in papers in *Nature* and in the *Journal of the Atmospheric Sciences* in 1978. More recent calculations continue to confirm them.

The year 1976 was magical for me because of my exciting summer at WHOI, and another trip that I will describe shortly. It was also the year of the blockbuster paper written by Hays, Imbrie, and Shackleton titled "Variations in the Earth's Orbit: Pacemaker of the Ice Ages" and published in the journal *Science* (Hays et al. 1976). In this paper, they had several 500,000-year records from two seafloor cores in the Southern Indian Ocean. I had the honor of meeting the authors at WHOI that very summer.

J. D. "Jim" Hays from Lamont-Doherty Geological Observatory, Columbia University, took on the recovery and analysis of one species of one kind of microscopic fossil skeletons (the top-dwelling, or *planktonic*, species) that indicated ocean temperature. Nicholas "Nick" Shackleton from Cambridge University figured out how to separate the water temperature signal from the total ice-volume signal, both being superimposed in the $^{18}O$ record. The temperature record was mainly due to the top dwelling (planktonic) foraminifera and the ice volume record was in the bottom-dwelling (benthic) foraminifera, the latter a rare foraminiferal species. The point is that the ocean temperature is nearly constant in the deep, revealing the ice volume signal, while very variable at the top, blurring the ice volume signal. Imbrie took on the statistical analysis. The authors took their data and analyzed it into contributions to the variance (along the core, which was equivalent to time) from different frequency registers—see the graphs on the right-hand side in the schematic diagram in the figure.

The authors of the blockbuster paper did not try to explain exactly how the orbital element changes brought about the waxing and waning of the great continental ice sheets. They merely showed that the timing and temporal correlation were right. In fact, to this day we still do not know just exactly what was the mechanism governing the great North American ice sheet expansions and collapses in these funny cycles.

In an interview with Imbrie recorded by the American Institute of Physics, remembering a presentation by Nick Shackleton:

> It's like a fingerprint. And you could recognize stage 1, 2, 3, 4, 5, 6, 7 in the Indian Ocean, in the Mediterranean, even in the South Atlantic, the Antarctic Ocean. And you can say this core here, whoops stage 7 is missing. There's a nonconformity. Or this core begins with stage 8. So suddenly, and so when he gave this record, there was a simultaneous cheer. Everybody in the room realized my God, we've got it!

At the same meeting, Neil Opdyke was in the room, and he had a magnetic stratigraphy (layer-by-layer analysis along the core) of the same core. In that record was the very strong and ubiquitously observed Brunhes-Matuyama (B-M) boundary, the last major reversal of the geomagnetic field, easily observed from locations all over the world and thereby fixing a benchmark at 781,000 years ago, way past the time that could be inferred from radiocarbon dating. The individual bumps and valleys in the ice volume versus time curve were given labels, stage 1, 2, 3, and so on. The B-M boundary was at stage 19. By making a linear assumption about the sedimentation rate, one could associate a time with each of the stages. Sedimentation rates are different from place to place on the seabed, but these can, at last, be reconciled. Although many participated, Nick Shackleton was the hero of this story.

The Hays, Imbrie, and Shackleton paper of 1976 was the "smoking gun" that pretty much sewed up the Milankovitch theory of the ice ages. I especially met and talked with Imbrie. I think he and Broecker might have attended my three lectures. They both were taken with the idea of a simplified climate model like the one I was peddling in my lectures. I thought there might be a prospect for my model, enhanced to two dimensions over the planetary surface and including the seasonal cycle, to be used as a good way to study the glaciations.

After I returned to UMSL from Woods Hole, we had an opening for a geologist who was to be housed in the physics department. Outstanding applicants came in because, as with my own experience in the field of physics, the job market across all of science had plummeted. We had candidates from CalTech and other great schools—not bad for a branch campus of a fledgling "red brick" public university. One candidate caught my attention, Tom Crowley, a recent PhD from Brown University and a dissertation student of my recent acquaintance and hero, John Imbrie. Tom had graduated and then took a civilian job with the Navy crossing the Pacific, teaching oceanography onboard. He was "burned out" by science (actually looking at those tiny critters and sorting them under a microscope). Tom arrived at UMSL for the fall semester of 1978. We will meet him in later chapters as he became a major influence in my career.

*Microscopic view of a variety of small life forms found in ocean cores. Picture attributed to PSAMMOPHILE.*

*IODP staff members examining a deep sea core. Note the old MacIntosh.*

QUATERNARY CLIMATES

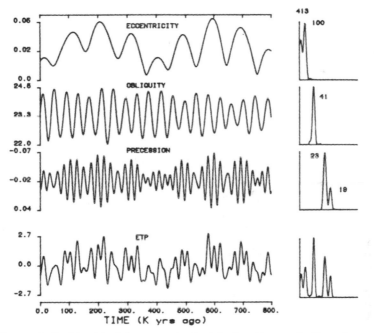

*Changes in Earth's orbital elements going back in time (thousands of years).*

# CHAPTER 5

# TASHKENT AND GLACIERS

I had the honor of joining a delegation of American climate scientists in the fall of 1976. We were to present and learn about the status of climate models in the United States and the Soviet Union. While I was there, the Soviets took us to a mountain glacier of interest in climate science, a very exciting side trip. Here, I take some time to discuss the role of glaciers in climate science.

## Getting to Abramov Glacier

While at Woods Hole I received a phone call, asking me if I would be available to join a delegation going to the Soviet Union (USSR) in late September 1976, just a few months away. This visit was to start in Moscow then on to Tashkent (in Uzbekistan, then a part of the Soviet Union, now an independent country) and last about fifteen days. I said I would have to get someone to cover my classes at UMSL, but it should not be a problem, and indeed it was not—thanks to my wonderful colleagues at UMSL.

I wondered how in the world I was selected. I think that a more important scientist than I must have canceled, and probably after several subsequent rejections, I received the invitation. Obviously, someone at NCAR or WHOI had recommended me. The delegation was to be led by Dr. Joseph (Joe or "Smag") Smagorinsky, the director of GFDL, located on the campus of Princeton University, a research institution within NOAA. Members of GFDL were civil servants and some had academic connections to the university. Smag was one of the first persons to use the new computer that was located in the town of Princeton at the Institute for Advanced Study. Knowing the territory, he later created the NOAA program in Princeton. He told me he wanted to get far outside of DC, because the bureaucracy and pressure from politicians would try to interfere with his research plans. He was a tough but wonderful leader, and I learned lots from him in this and later encounters.

We stayed in the three-thousand-room Rossiya Hotel, built in the 1960s by the Soviets. At one time it was the largest hotel in the world. Just a year after our visit, the hotel suffered a fire in which forty-two persons were killed. It was demolished in 2006. It was a short walk to the spectacular Saint Basil's Cathedral in Red Square and the adjacent Kremlin. The Cathedral was built in 1561 by order of Ivan the Terrible after his defeat of the Tatars in the city of Kazan.

The delegation had a number of important scientists interested in the climate problem. Besides myself, the group included Peter Stone of MIT, Yale Mintz of UCLA, Lawrence Gates of Oregon State, Jong Yi Kim of Oregon State, Barry Saltzmann of Yale, Sukuro Manabe of GFDL, Ruth Reck of General Motors Research Lab, and Smag's fifteen-year-old son, who was studying Russian. There might have been a spoken language interpreter and an administrative person from NOAA as well to assist us (I do not remember). I had met Saltzmann and Stone at Woods Hole—both of them had read my early papers and later served as references for my promotion to full professor of physics at UMSL.

On our first morning, we were taken to a high-level office, probably at the Soviet Academy of Sciences. We met in a room with highly polished furniture, a large oblong table with flowers and a supply of bottled bubbly water at the center. In our briefings by US officials before arrival, we were warned not to drink any water from a faucet—not even to brush our teeth (the intestinal parasite, giardia, can take months to cure). Everyone in the delegation knew one another, except for me, and Smag's son. I knew no one, and only two knew me. Each of them was kind and cordial with me, far more than I expected. Superstars in meteorology seemed less arrogant and condescending than their counterparts in theoretical physics, wherein hierarchic standing would be quickly determined and acted upon.

Through the door of the conference room came an elderly, fierce-looking, well-dressed, academician, Evgeny Fedorov, a glaciologist, unknown to me, of course. Fedorov was not only a distinguished glaciologist, but he headed an institute and even founded the *Applied Geophysics Institute of the Soviet Weather Service*.

Becoming an academician in the Soviet Union was not a trivial achievement, but of course, politics was always a factor to gain entry and gain further recognition. Even so, the Academy of Sciences was somewhat less political than some sectors of the Soviet system. Highly accomplished scientists had more leeway than most people with the same level of achievement, such as managers. Members of the Academy had access to special stores for groceries and other merchandise, sometimes only accessible to tourists for hard currency. They often had dachas and finer apartments and sometimes cars—sometimes even chauffeurs.

Fedorov said forcefully, "There is a recent CIA report showing that increases in $CO_2$ [carbon dioxide] are causing climate change, and it could be a serious threat to society. Does the American Delegation agree with this assertion?" Remember, it was 1976 and climate change due to $CO_2$ or the Sun's changing brightness were not exactly a topic of dinner conversation in America. We looked at each other puzzled, not knowing what to say. I certainly, and I suppose none of the Americans had even heard of such a CIA report either. It made me wonder who got US scientific information first, the Soviets or the Americans. He then said, "Some others are saying that the Sun's brightness is changing and instead that is the problem. So which is it?" We essentially told him that some scientists have suggested that $CO_2$ might be a problem, but the science is simply not settled at this time. But American and Soviet scientists certainly should consider these issues. It was an area with common ground for cooperation.

After this meeting we went on a tour of a prerevolutionary estate with beautiful gardens in Moscow, the Arkhangelskoye Estate Museum, the complex of grounds and buildings were planned and implemented for a Prince Golitsyn in the late eighteenth century by a French architect. In 1810 the property was bought by Prince Yusupov, a wealthy descendant of the Tatar Khans. I mention this because a member and descendant of the (rather large) Golitsyn family became a personal friend and appears later in my story.

Because I ended up making a total of five visits to the USSR in this exchange program, I will spend a few words to explain how these visits came about. I knew virtually nothing about it at the time, and never did fully understand it, but it goes something like this. President Nixon and Soviet President Nicolai Podgorny signed an agreement in 1972 (Nixon was in the middle of a hot reelection campaign as well as the Watergate scandal).

President Nixon and Communist Party Secretary Leonid Brezhnev (where the power was concentrated), both normally thought of as hard-liners, took some small steps to thaw the Cold War. Brezhnev followed Nikita Khrushchev, who was removed from office in 1964.

In the agreement, there were a series of working groups, each to consider one area of the environment. In particular, our program, Working Group VIII, was devoted to pollution and climate change (see Bierly and Mirabito 1984). I do not know much about the missions of the other working groups, but Working Group VIII promoted visits, exchanges, and conferences held jointly between the two countries. Our visit was one of the first in 1976. John Imbrie noted during an interview conducted by the American Institute of Physics that he attended such a meeting in Baku, Azerbaijan (then a republic in the USSR, now an independent country) during roughly the same time interval. The Baku meeting focused on paleoclimate. The Tashkent meeting we were to attend was focused on climate modeling.

From Moscow, we flew on the austere Aeroflot Airlines to Tashkent, the fourth largest city in the USSR at that time. The region is fairly dry with only about 43 cm (seventeen inches) of rain, annually. The climate is continental with hot summers and cold winters. The weather in Tashkent was pleasantly mild during our visit. This meant that in the late afternoons there was an opportunity to walk the pleasant streets with my new friends, looking for restaurants. Actually, it was a bit like a camping trip. We were able to get to know each other well. I enjoyed the food, especially the bread that was baked in a pie-like disk about an inch deep and brown on top. You simply tore off a piece and ate it by hand. Also, there were fresh fruits and melons (one should never eat fruit without peeling in the USSR). This seemed to be quite a change from Moscow where the sample of food we encountered was generally lacking.

The lectures were generally thirty to forty-five minutes long, with interpreters. We wore somewhat uncomfortable, raspy earphones into which were injected the voices of the several Russian speaking interpreters. The exclusively female interpreters were friendly to the Americans, each interviewing us before we presented our talks, in order to get a feeling for our speech characteristics. There was no PowerPoint in those days. We had prepared in advance thirty-five-millimeter slides. One could also use an overhead projector with transparencies. There were the usual universal calamities such as bulbs failing. In truth, the Soviet lectures were very hard to understand partly because of the different presentation styles of the Soviet lecturers. Graphics on the screen helped, but even they were hard to see. It was clear though that if they were representative, the Soviets were far behind the United States in terms of progress in meteorology and the emerging climate science, especially its implementation on computers. Their stars were as good as ours, but there were few of them among a large number of atmospheric scientists. Their research scientists simply did not have much access to high-speed computers, or even copying machines.

I learned the most from the American lectures. I needed to acquire more understanding of general circulation models, and these were the people to learn from. Not only did I learn in the lectures, but I was also able to ask questions of them privately. I learned, and they found out that I was serious and soaking it up.

The Soviets excelled in a few categories. Many of these seem to have derived from one of the most influential mathematicians of all time, Andrey Kolmogorov. Kolmogorov, whose life nearly spanned the twentieth century, was a leading figure in statistical mathematics. He made contributions in theoretical physics with his extraordinary work in power spectra of turbulence (how the kinetic energy in turbulent flows is distributed over geometrical scales: mostly in large scales, diminishing toward smaller ones: he had a formula). He also worked in probability theory, classical mechanics, topology, and analysis. He was an advisor to such well-known figures as Vladimir Arnold, Israil Gelfand, Andrei Monin, Alexander Obukhov, and Akiva Yaglom. I have read books by Arnold and Yaglom, in English translation, of course. Kolmogorov's influence was in many areas of mathematics (turbulence, stochastic processes, and others) that are relevant to meteorology and climate science. We will encounter two of his most important students related to climate science, Andrei Monin, an oceanographer, and Alexander Obukhov, an atmospheric physicist, later in this book.

One charismatic young scientist was Sergej Zilitinkevich. He presented his work on the mixed layer of the ocean and its relation to climate. The mixed layer is an important feature in the world ocean. The mixing is due to the wind rubbing against the ocean's surface coupled with larger-scale currents of the ocean. The wind at the surface causes waves that break and also transfer kinetic energy below from the atmosphere. This layer is well-mixed because it is constantly stirred by these interactions above and below. It is variable from place to place and the layer ranges from 50 to 120 meters in thickness, again depending on season and location. One reason it is an important indicator of climate is that when the heat balance is changed, the amount of mass that changes temperature is partially determined by the thickness of the mixed layer. Thermally, the mixed layer is approximately isolated from the water below, so the heat capacity of the water column depends on the amount of mass above the bottom of the mixed layer. Understanding this part of the ocean is very important in hooking it to the atmosphere in a climate model.

Zilitinkevich had an elegant theory of how this phenomenon worked. But as a person, he was very busy flitting about from one of us to another—I was an unknown to him and was treated accordingly (i.e., ignored). Some in the US delegation said they thought he was involved in schemes that could get him in trouble with the authorities. Rumors suggested that he did get into legal trouble later. I never heard of him again until I saw his name on a poster about a seminar to be presented in the Department of Physics at Texas A&M in 2015 or 2016. It had been forty years since I had seen him so it was possible I might not recognize him. It turns out it was the same fellow, and I had a nice chat with him after the seminar. He now works for an office in the weather service in Helsinki, and he is still poking away at turbulence theory—one of nature's most difficult to treat. I was wisely steered away from working on turbulence theory in my sabbatical at NCAR. Some describe turbulence theory as a black hole sucking away talented people.

John Imbrie had an earlier visit to the USSR than my first. In his visit to the Working Group VIII conference in Baku, his Russian host, Gerasimov, asked what would be the most important thing he would recommend to strengthen research in paleoclimatological research in the USSR. Imbrie replied that they should buy and distribute a lot of Xerox machines. Apparently, they had copying machines, but one could not copy a page without filling out forms and having them authorized by someone well above their pay (and probably political) grade. I think he was frustrated and confused in the USSR by its extreme contrast with the way we do things. Photocopy machines were a threat to the regime, because they could be used very efficiently (compared to typed carbon copies) in underground dispersal of unauthorized information—most famously, novels.

Even though things could be exasperating at times, our Russian colleagues extended kind hospitality toward us. There were the usual big dinners with many courses, vodka, and toasts. It was hard to stay moderately sober with all the glass-filling coming from tipping bottles over our shoulders by waiters with white towels drooped over their forearms. I developed a taste for caviar—OK, vodka, too.

Mikhail I. Budyko and I made acquaintance in Tashkent at last. He was one of the several hosts at the conference. I could never tell at that time exactly who was running things on their side. He was a tall and handsome sixty-year-old with a smooth manner. He obviously was a very cultured man, and I found even more to admire about him during my trip the following year. His English was excellent. I will always remember how he began the conference with a large portable blackboard behind him. He had a piece of white chalk that looked as thick as a peeled banana. He stepped to the board and wrote that the change in global average temperature for a 1 percent change in the Sun's brightness would be about 1.0°C (or 1.8°F). This was based upon his very simple model of the climate system. He even acknowledged that my paper gave a similar result. The same result comes from calculations made by a more sophisticated climate model, as recently published by Dr. Sukuro Manabe (in the audience). He also alluded to the unknown value one might expect if the concentration of $CO_2$ was doubled. The mystery underlying these calculations lay mainly in the feedback mechanisms in the atmosphere. The feedback mechanisms should be comparable for either the solar or the $CO_2$ disturbance. It was thrilling for me as the talks continued throughout the day. Budyko was my new hero!

Budyko was especially interested in me because my work extended and clarified the role of his in the model hierarchy. I was careful in my 1975 papers to praise his work. He took me aside a number of times to discuss our mutual interests. He introduced me to his colleague Lev Gandin, with whom I would form a lasting friendship. As the conference went on, I spent lots of time with Gandin, and we began to speak of my making an extended visit to Leningrad (now St. Petersburg) the following summer (1977). The Working Group VIII leadership were able to agree on such a visit, and I will describe it later in the next chapter. There were perhaps fifty Soviet scientists at the conference, including Georgii Golitsyn, Igor Karol, Ilya Polyak, and many others, who will also appear later.

## Glacier Science

We were in Tashkent spanning two weekends. Soviet hospitality included excursions on each. The first was a trip to Abramov Glacier (39° 38' 55" N, 71° 35' 09" E). This is a famous mountain glacier in Kirgizstan, very close to the border of Tajikistan. We flew from Tashkent to a small airport probably in Kirgizstan, where the party included Budyko and Gennadii "Guena" Menzhulin, one of his assistants, whom I had met earlier in Moscow. I think there were only a few other Russians along. We boarded two large Soviet military helicopters. The choppers took us up into the mountains just across from China and Afghanistan. The altitude was over four thousand meters (twelve thousand feet). We jumped out, suiting up in the provided boots, coats, and old-fashioned Russian ear-flapped fur hats, then strolled down a trail with the glacier sprawling to our left and right like a great river down the slope in front of us. A few, including Prof. Mintz, rode donkey-back. There was a research station a few hundred meters above the glacier's edge. I had never seen a glacier that grand. I had seen some small ones in Glacier National Park in Montana during a camping trip with Jane in the early sixties, but nothing like this. It was nearly as wide as the Mississippi River, rumpled on its surface with lots of dirt and other debris. The debris came from the glacier's edges scratching up rocks from its terminus, scraping lateral edges or dislodging and scooping hunks of bedrock below. We took a walk up and along a moraine (mass of rubble and rocks piled up about fifty or more meters alongside the glacier). We crunched alongside the glacier on a path that was partially cleared of snow probably just for our visit. After walking about a kilometer (0.7 miles), we decided to turn back. Prof. Saltzmann was near passing out from the altitude. We feared that he had altitude sickness. He soon recovered after we entered a small cottage where there was an incredible dinner laid out for us. It was nice to have some hot tea, but after that and given the altitude, I limited myself to only one shot of vodka—uh, maybe two.

There was and still is considerable interest in the "health" of glaciers. At that time, scientists were concerned that they might be shrinking. (Of course, now it is certain.) Hence, the scientists at the research station were taking continuous measurements of the precipitation, temperature, and the conditions at the surface of the glacier, basically its mass balance—how much water in the form of ice it gains from precipitation, how much it loses from melting and evaporation. A glacier flows downhill, usually terminating into a regular river, or in some cases, it flows all the way to the sea. Some mountain glaciers terminate in a lake.

Increases above average in the rate of flow of ice from an ice sheet contribute to a rise in sea level. Such ice may have been stored on the land for thousands of years. This is the case for the great ice sheets in Greenland and Antarctica. Both are delivering more fresh water in the form of ice to the sea than there is fresh water in the form of snow and ice being delivered by precipitation over the ice sheets. The processes for these transfers are very complicated, mostly because the physics of glacial flow is still not fully understood. The mechanics of the flow of ice are much more difficult than the flow of water in the ocean currents, or flows of air in the atmosphere. The devil is in the details of how the forces of friction or viscosity are expressed in Newton's laws of flow for glacial ice. These new terms in the flow equations introduce more unfamiliar effects than in the simpler cases of water and air.

Finally, we are late in getting started in the world of modeling glacier mechanics. We simply have not been at it as long and with sufficient zeal compared say to weather forecasting, where we have now half a century of a dense global network of hourly observations and daily weather forecasts. These forecasts have to face the music (public scrutiny) day-in and day-out. This regimen forces the whole forecasting enterprise, from data gathering to modeling, to improve steadily or else close the shop. Oceanography has entered this phase of periodic user scrutiny, but the forecasting phase of glacial flow is decades behind. Fortunately, experience in the sister fields, supercomputers, and now dedicated satellite missions are increasing our understanding of ice flows in nature.

## Samarkand

Let us return to my first trip to the USSR in the fall of 1976. After the weekend in the high mountains of Uzbekistan and Kirgizstan, we were exhausted, but on Monday morning we had to resume in the auditorium in Tashkent, starting with those pesky earphones. Maybe US earphones would be just as uncomfortable back home, but I just have not had that experience. We spent the week mostly listening to talks given by our Soviet colleagues. In his talk, Prof. Yale Mintz from UCLA delivered a condescending contrast of the American and Soviet work in the world of big-time climate modeling. Yale had been a pioneer in GCMs beginning in the early 1960s, first for weather forecasting. UCLA already had a fine Japanese meteorologist Michio Yanai. When Mintz recruited Akio Arakawa, Yanai refused to be promoted to full professor until Arakawa was—a Japanese gesture of respect. All three made UCLA a great department in the field of numerical weather and climate modeling. Mintz was indispensable not only in the science and recruiting but in generating the financial resources for accommodating large-scale numerical modeling.

We were happy to see the next weekend arrive. This time we boarded a short local flight to the legendary city of Samarkand, also in Uzbekistan. Samarkand is slightly off the old Silk Road joining China on one end and Europe on the other. The road may have been open and traveled as long as two millennia. It was an active trade route in the Middle Ages. To Westerners, the Venetian Marco Polo is probably the most famous of the travelers along that overland route to China.

In the early Middle Ages, the great Mongolian conqueror Genghis Kahn took over control of the Silk Road. He was a brutal warrior, who typically killed every man in a village or town as he took it, and while he was at it, he took the most beautiful women as his wives. However, he gets some credit for putting the Silk Road under some kind of administrative umbrella.

The architecture in Samarkand is stunning. In 1976 the magnificent buildings were not in good repair as those in the recent photos. In the 1970s lots of weeds and brush were growing atop those brilliant, but dirty turquoise domes; some appeared to be giant bird nests. The colors of the tiles of the mosques and madrassas were breathtaking, especially considering the drab colors of the Stalinist buildings in Tashkent.

One incident in particular from my visit to Samarkand is worth relating. Several Americans and I, along with Guena Menzhulin, Budyko's assistant, went into a rather bleak restaurant for lunch. We were growing irritated after waiting for a long time with no service at our table. Guena checked around and claimed that they thought we were Russians. Guena was of Georgian descent. He explained that we, fair-complexioned guys, were from the Baltic Republics, and that made us OK (Americans might not be so well-favored even as Russians). We were served sparkling water in dark green but transparent bottles similar to some dark green beer bottles in the West. I could see residue at the bottom of mine. It swirled about when I tipped the bottle. Further examination showed that the residue was the fragmented remains of some kind of creature that looked like a two-inch grasshopper. They brought me another bottle, but I did not drink from it.

A few days later we were on our way back through the airport in Moscow. Well before the trip to the USSR, Larry Gates had suggested that it might be a good idea to stop over for a night or two in London on our way back. I had agreed, and booked a room in a bed and breakfast in London. This was my very first trip to Europe so I was curious. All alone, I had a fine time sitting atop the two-story red tour bus, traveling around London and visiting some museums and Westminster Abby. I was beginning to feel like a well-traveled *bon vivant*.

## Return to St. Louis

As I mentioned in the introduction, my sabbatical trip to Boulder had put an enormous strain on my marriage to Jane, who did not accompany me for the year in Boulder. By early 1976 we were formally separated, and with the unhelpful assistance of marriage counselors, we were headed for an amicable divorce. My next few years in St. Louis were partly work, but mostly misery.

I spent one year living in and minding the house of my friend Phil James, who took his family on sabbatical to JPL in Pasadena, California, for the whole year. Phil was on a similar path to retread himself to become an expert on the ice caps and other features of Mars, fully understanding that high-energy theoretical physics was not for either of us. His minor had been in astronomy for his PhD, in contrast to mine in math. When he returned to St. Louis, we wrote a paper applying a variant of my energy balance climate model titled "The Seasonal $CO_2$ Cycle on Mars— An Application of an Energy-Balance Climate Model," published in the *Journal of Geophysical Research* (James and North 1982). The paper has more than seventy-five citations.

I need to say a few words about my old friend Phil. First, he is an extraordinarily bright guy who was interested in everything (e.g., European history). I think he could solve mathematical puzzles (e.g., integrals in calculus) faster than anyone I know. He combines his quick mind with an uncanny memory. He met and married his sweetheart Sharon while in graduate school. We played many a bridge game with them and other couples on weekends. He saved me from professional penury by suggesting me for the faculty position at UMSL. He is honest and a very good scientist, editor, and teacher. After I left UMSL, he went on to be the department head of physics there and the same later at the University of Toledo, where he also became a University Distinguished Professor because of his Mars work. He also served for many years as editor of the *American Journal of Physics*, a periodical devoted mostly to college and university physics teachers. It is a journal dedicated to historically interesting, but also fresh ideas and curious problem solutions in physics. It has been around forever and is highly regarded for its quality by the physics community. Phil was a master at dealing with authors (mostly cranks) whose goal was to prove that the mysterious laws of thermodynamics or relativity were wrong—for example, perpetual motion machines. Some of these puzzling attempts to uproot established theory are very knotty, but Phil could always find their flaw and gently disabuse the author of such nonsense. As an expert on Mars and especially its polar caps, he had continuous funding from NASA even into retirement.

I had never met Marvin Geller, who was spending a year at the University of Miami from his regular faculty position at the University of Illinois at Champaign-Urbana. Marvin had heard of my work and suggested that Miami take a look at me, since they had an opening. The department head, Jack Geisler, called me about visiting for an interview. I took the trip and spent extra time with oceanography professors Eric Krauss and Claes Rooth, the latter of whom I had met, admired, and liked at the WHOI summer program discussed in the previous chapter. Marvin and his wife had me over for dinner at their house. At this point, Marvin had decided to stay at Miami and give up his professorship at Illinois. After my seminar, Krauss offered me the position at a sizable increase in salary from UMSL. When I told Krauss of my marriage dissolving and the problems I was having separating from my children, he told me that I looked terrible, slumping and smoking too much, for starters. He was right, of course. He said that it might be better to be in a different city from the kids. Thus when they are visiting, I could devote quality time to them, and they could not say, "I am bored, take me home to mom." This made some sense to me, but I was hardly sure about that idea.

I returned to St. Louis perplexed. Soon I received a call from the University of Illinois, wondering if I might be interested in a job there. Marvin, having given up his job there, had recommended me. Also, an insider at Illinois named Takashi Sasamori, whom I had known from my time at NCAR (his house was next door to the one I rented in Boulder), was also pushing for me. This seemed more interesting to me, since it was only about a four-hour drive from St. Louis. I drove to Champaign, gave a seminar, and met everyone. The department head there was Prof. Ogura, who was away on leave for the year. An interim head met with me, and we discussed the idea. I came back to St. Louis, thinking I would get the offer. One option was that I might use these offers to negotiate an increase in salary from UMSL in case I wanted to stay or didn't get the offer. After all, my dean had come to UMSL from a senior full professorship in biology at Illinois, and I know he thought highly of his old school, and so did I: Illinois is one of the great public universities. Unfortunately, I did not get the offer from Illinois. I sent a handwritten note to the interim department head at Champaign asking that he write me a letter saying how well I did in the contest—a dumb idea. He sent me a letter saying that I was highly qualified and was in the top ten applicants—hardly what I wanted to hear. So I took my offer from Miami to my dean and presented it wondering what he might do for me. Of course, he said, "Well, that looks like a good offer. It would be hard to turn it down." I slinked back to my office, rather deflated from my high opinion of myself. I just could not take the job at Miami at the time because of my family situation. It was just too far from St. Louis, the kids did not want to attend the University of Miami, and the school did not have a tuition arrangement with others. Moreover, I felt uncomfortable in Miami.

USSR climate scientist Mikhail Budyko (left) and Jerry (right) in Tashkent.

US delegates at a statue commemorating an earthquake in Tashkent. (left to right) Larry Gates, Smag's son, Ruth Reck, Joseph Smagarinsky, Barry Saltzmann, Jung Yi Kim, Yale Mintz, Peter Stone, and Suki Manabe.

*Jerry in Samarkand, 1976.*

*USSR military helicopter with USSR and US climate scientists near Abramov Glacier. Photo by author.*

US climate scientists walking the Morraine above Abramov Glacier, 1976.
Photo by author.

The Magisterial Entrance to the Main Geophysical Observatory in Leningrad
(now St. Petersburg).

*Meteorologist Lev Gandin as a young scientist at the Main Geophysical Observatory.*

*Tom Crowley in the early 1980s at a Chapman Conference in Colorado.*

# CHAPATER 6

# LENINGRAD SUMMER, 1977

In a side story, I spent most of a summer in the Soviet Union in 1977. The stories I relate give some idea of science, scientific institutions, and the life of scientists during the Brezhnev period. I tack on another trip, this time to Soviet Georgia. It includes more American Scientists and our adventures.

## Summer in Leningrad

During my trip to Tashkent in 1976, Lev Gandin and I solidified our friendship. He suggested that I return to the USSR under the Working Group VIII program for an extended visit. I thought a seven-week visit would be enough for me to do a little research, present some lectures and get to know more Russian scientists with whom I had mutual interests. An additional weeklong trip to Novosibirsk was tacked on. This turned out to be satisfactory with the authorities on both sides. Before leaving in 1976, I promised that if I should return, I would bring some gifts. Guena (diminutive for Gennadi, both pronounced as in "guess") needed some blue jeans (as did everyone else), and he gave me some measurements. I agreed to bring a few books and a record (The BeeGees) for interpreter Irena Mershchikova. I was able to do both as well as adding a few books for Gandin and Budyko.

I arrived in Leningrad late in the night. Guena met me at the airport, and we made our way via taxi to the Oktyaberskaya (translates to "October," for the 1917 revolutionary event in the city) Hotel, located just a few steps from Nevsky Prospect (the main street in downtown Leningrad) and the Vostania Metro Station. There is always a person sitting out front of such buildings on a chair all through the night to keep undesirables from entering the premises wherein foreigners might be contacted. This was usually an elderly, inebriated gentleman. When we arrived, he shouted at us, "Gyer-manin?" Guena said the gentleman wanted to know if I was a German—hatred left over from WWII, which was especially bitter for locals because the great city was under siege for nearly three years. Guena repeated the technique he had used before at the lunch table in Samarkand. In the presence of local folk, he always explained that I was from the Baltic countries. In the Asian republics, it was to distinguish me from Russians, who were apparently not so well-liked. In Russia, it distinguished me from Germans, and among Russians, it helped explain that I could not speak Russian with any ease, but that I was from a friendly Republic.

The next morning Victor "Vitya" Frolkis, a graduate student who was to become my guide, friend, and companion at the Main Geophysical Observatory (MGO), where my office was located, picked me up. You will find more about him in my little (true) story called "The Key" that appears as a separate section later in this chapter. The MGO was an old but distinguished looking building, but a bit crumbly inside. The MGO was founded in 1849 and actually still carries the founder's name, Viokov (MGO is sometimes referred to as VMGO).

On that first day, Igor Karol, a meteorologist specializing in the stratosphere, perhaps a few years older than I, proudly pointed to a portrait of perhaps the most famous scientist who ever worked at MGO. This was Alexander A. Friedman, who found some extensions to Einstein's general relativity equations that allowed for solutions permitting an expanding universe. In the textbooks I had used in the United States, it was referred to as the Friedman Universe. Friedman had received his doctorate at St. Petersburg (Leningrad's name before the Bolshevik Revolution) University in 1910, and afterward, he taught at a mining institute. He was a pilot in WWI, later becoming head of an aircraft factory. He published his famous paper in 1924. He was appointed director of the MGO in 1925. Sadly, he died that same year of an illness, possibly typhoid fever, contracted in the Crimea. He was doctoral supervisor to George Gamov, a legendary and colorful figure in the early days of quantum theory, later to immigrate to the United States and teach at the University of Colorado. As an undergrad, I had read Gamov's popular science book *One, Two, Three, Infinity*, given to me by Jane's older brother, Bill. As in all of his popular books, Gamov's drawings and sense of humor were fun, the kind of fun I saw in many Russians. I was inspired by Gamov's books in my early twenties.

Mikhail I. Budyko had held the directorship of MGO for eighteen years, until just before my trip, when he was apparently deposed from that position, likely for political reasons—my opinion, based on rumors I heard. Being fast on his feet in the Soviet bureaucracy, he became head of a branch of the State Hydrological Institute (SHI), itself an extension of the Hydrometeorological Center in Moscow (similar to our NOAA). Ironically, the SHI was located in the MGO building, the very building he formerly ruled over. The new director of the MGO when I was there was a man named Evgeniy P. Borisenkov.

I got the feeling that Dr. Borisenkov was more in line with Mr. Breznev than was Prof. Budyko. But the Soviet system had many channels up to the top and back, and it is difficult for an outsider to navigate them.

## The Key

Next is a short and I hope amusing story I wrote immediately after my visit to Leningrad in 1977—every word is true.

I spent eight enjoyable weeks in the USSR this past summer (1977) doing research and lecturing on simple climate models. I spent seven weeks of the time in Leningrad and a week in Novosibirsk (in Siberia) and a few days in Moscow. I have recorded a few curious impressions, which might amuse the occasional traveler.

My first day at the institute was filled with much fanfare. I met the director and all of his immediate subordinates, and was finally shown my office and given the key. The institute is 125 years old, and it seemed like everything was preserved from the original structure, including the plumbing. I occupied a dandy office on the first floor all to myself. I was told that all foreign specialists get to use this office. With the help of my pocket Russian-English dictionary, I discovered that the office next door was labeled "party bureau." Outside of my office on the walls hung the usual pictures of Lenin and a whole bulletin board devoted to pictures of Secretary Brezhnev and members of the Politburo—this latter vanished suddenly the next week when President Podgorny was removed from office; it was replaced with news about the latest evils of Zionism.

They explained to me that I shouldn't lose the key because no other existed. I promised to be careful. One of the younger scientists—"Little Sasha," as I called him—was assigned to be my guide. This fellow wanted to show me every sight in Leningrad. He could walk ten miles in a driving rain just to show me one more monument. The first problem was how I was to get to the office each day. It was too far to walk, and I think they did not want me wandering around on my own, perhaps I might get lost or get in trouble. One suggestion was that an institute driver would come and get me each morning. The cars were unreliable, but I could do it this way until they all broke down (seriously). I insisted on riding the subway and Little Sasha showed me how—I suppose he had to get high-level permission. It involved one change of line followed by either a one-mile walk or a bus ride—I chose the former since buses at rush hour are another story. Little Sasha insisted on coming to get me for a few mornings, since this way he could arrive at work an hour and a half late. I managed, at last, to convince them that I could find my way to the office each morning.

But if I wanted to attend a ballet or opera in the evening, it always meant being an hour and a half late to the office anyway. The way to get tickets was interesting. A booth in the hotel was open each morning from 9 to 11 a.m. You first got pieces of official paper from the booth. Then you took these to the hotel cashier to pay for them. She signed each of the documents, and you went back to the line at the first booth. You turned in the signed documents and were told that if you returned in the afternoon between 3 and 6 p.m. you could have coupons that you could exchange at the theater box office for tickets. One time, I returned after 6 p.m. and my circus tickets were sold to someone else. The ticket lady learned some new English expressions that afternoon. (Much as I desired, I never got to see the Moscow Circus on this trip.)

On the first Friday after my arrival, Little Sasha wanted to see me before I left the institute. He wanted my key. He said it was the only key to the door, and they were afraid I might lose it over the weekend. I was much more afraid that they might lose it after my week's experience. I tried in vain to explain to him in broken Russian-English that there is another key because someone comes in every night and empties my ashtray. I wanted my key for two reasons: all of my work was in the office, but worse yet, I would have to wait for Little Sasha on Monday morning. In the ensuing struggle he got the key, and on Monday morning he showed up red-eyed and half an hour late to take me to the observatory because only he could retrieve the key for me.

On the next Monday I went in by myself, and when I reached the institute, I wanted to call Little Sasha to bring me the key but realized that my phone was locked in my office! No one had shown me where their offices were located, so I began wandering through the building (highly illegal for a foreigner, I learned later). Finally, I found a very distinguished professor I knew and told him the problem. He found another very distinguished professor who after ten minutes of investigation found out that Little Sasha was home, sick. Both professors began combing the building for my key. They discovered that the key room was locked. At last, the room was opened, and I had my key at 11 a.m.

Each day was filled with conversation at coffee breaks—provided coffee could be found in the city of Leningrad that week (special efforts by my friends were made to get really good coffee with me in mind). I was the center of attention because of my speaking English, and being a Southerner, I spoke slowly and refrained as much as possible from tricky idioms. A crowd always gathered with scientists, interpreters, and graduate students. We talked endlessly about America and its customs and special language. Most of them could communicate in English, and they loved to practice when they got hold of a native English speaker. After lunch, a few always came to my office to smoke American cigarettes, tell jokes, and shoot the bull. Their favorite question was "How much money do you earn?" Finally, I had to speak about the key.

(It was always assumed that my office was bugged. Occasionally, this entailed manual expressions such as a finger across the lips, or a finger pointed at the ceiling.)

At last I learned that everyone has to leave his/her key on a pegboard in a special room every night on the way out, and that it would be "too much trouble for me." I asked Little Sasha what if he wanted to work on a Saturday or Sunday or for that matter after 5:30 p.m. He explained that it was easier to work in his apartment anyway. Indeed, I soon learned that my best work was done on Saturday and Sunday at the hotel. In reality, I was quite busy on weekends, saturated with offers to tour Leningrad and its incredible environs.

My first hotel actually had a desk in the room. By the end of the first week, I noticed that I had company in the room. Bedbugs. These were a hungry lot, and when I stopped at the US Consulate (near the famous Finlanskiya [Finland] Station where Lenin reentered Russia to take charge of the Revolution) to pick up my mail (and read the *New York Times*), I mentioned it to the secretary there. She said, yes, that hotel was notorious for them, but the manager would deny their existence. He did. I moved to another room—worse. By this time, I had about forty bites so I spoke to the powers at the institute. They said it was impossible; after all, they didn't have them at home. It must be mosquitos. I suggested that maybe I should move on to Novosibirsk six weeks earlier than planned. That afternoon there was a knock at the door, and I was moved to new sleeping quarters.

My first hotel on this trip was the *Oktyeberskaya*, then home of bedbugs. My second was the exquisite prerevolutionary *Yevropeeskia* (Hotel Europe, which now is a famous, remodeled luxury hotel), just a block off Nevsky Prospect, and in the opposite direction, a block from the Russian Museum of Art.

After my experience in the first hotel, I learned that Leningrad is famous for its bedbugs and that the opera house and subway seats used to be crawling with them. The timing of the bites was said to be linked to the events in the program. Someone at one of the coffee breaks even conjectured that bedbugs were an endangered species, and that Leningrad had been set up as a sanctuary. I can only attest that they like imported meat.

Leningrad is beautiful in mid-June. It is so near the Arctic Circle that for several weeks in this period the sky stays lit all night. It is called the time of "white nights," and several cultural festivals are held during this period. My favorite thing to do was to go out on one of these nights with a group of friends. The process was to first consume a bottle of vodka, then take the last subway to the center of town at midnight. The Neva River runs through the city, and many ships go up and down the river. All of the bridges are drawbridges, and they open at 2 a.m. to let the ships pass. They then stay open until 5 a.m. So if you get caught on the wrong side from home, you had better have a blanket. We walked all night among the crowds and finally stopped at a friend's place for tea, cheese, and grain alcohol mixed with water. It was then I think that someone explained that the room for keys was just around the corner from my office, but foreign scientists were not allowed to go in there. It made perfectly good sense to me because all foreign persons are to be suspected, and to this day I can't understand why they did not tell me that on the first day. I suppose they might have thought that I would be offended.

My last week in Leningrad coincided with a joint US–USSR symposium, and I was invited to participate, so that week I was without need of the key except on Friday when I wanted to go into the institute one last time to clean out my desk. Little Sasha went in with me, and while I waited outside the office, he ran for the key. When he returned, we tried to open the door, but the lock would not work. It seemed to me he had gotten the wrong key. At any rate, I tried again to explain to him that there was another key that was held by the janitor or some other nocturnal creature with a clean ashtray compulsion. He couldn't understand me. He also couldn't understand the possibility that the lock had been changed. He went for help. Soon he and another young man returned with some tools. At first, the maintenance man tapped gently at the lock. When Little Sasha disappeared, this guy began to get more violent wedging a chisel in the crack in the door and apparently trying to break it down. Suddenly, the door flew open and Little Sasha was standing inside. He had crawled in through the window. Turned out that with all the fuss about the key, one only needed a penknife to unlock the window. Later, he returned, showed me how when you locked the door you must only turn it one revolution, not two. Rather than argue, I nodded and left for Novosibirsk.

The Soviet Union was a vast country. We traveled all day, including a change of planes and airports in Moscow. They sent along another young graduate student to keep me safe along the way—Little Serozhia. The trip to Siberia crossed four time zones, and most of the country under us was covered with forest. I couldn't believe it. There was a great paper shortage in the Soviet Union—that includes toilet paper. It was impossible to find good books in the shops. I wanted some of the classic Russian literature partly as a souvenir and partly as an incentive to continue my study of the language. It was not to be found—not because of prohibitions but because of paper shortages. There were no paper bags at the grocery stores—shoppers brought their own bags made of webbed string. Usually the cafes did not have napkins, and when they did, they had to be cut into three-by-three-inch squares   a problem if you had a mustache and you were addicted to Russian ice cream. The problem became so acute that long lines would form if some store was found to have toilet paper. The limit was two rolls per customer. I never had a supply problem at the hotel, but in the institute restrooms, I noticed that *Pravda* was being used for more than one purpose.

Most Soviet paper it seems is imported from Finland. I asked one person in Novosibirsk about it. He said that the Soviet factories were rather inefficient and that most of the logs are cut and floated down the rivers, but many of these get away into the Baltic Sea. There is a circulation pattern in the Baltic that then carries them to Finland, and the Fins retrieve and make good use of every splinter. I think the story might have been exaggerated.

While in Novosibirsk we learned of some mix-up in our travel plans to Moscow, so it was necessary for Little Serozhia to make a number of phone calls to Moscow and Leningrad. We gave him courage, and he pounded his fist until the problem was finally solved. But while on the phone to Leningrad, he stopped and said the authorities in Leningrad could not find the key to my office—where was it? I said I had given it to Little Sasha. They said Little Sasha had gone on vacation for a month to the Black Sea. But before I could explain through Little Serozhia that either the lock was broken or changed, or about the penknife and the window, or the nocturnal creature, they had hung up.

Leningrad is now St. Petersburg. The story above is completely true, except that the names of the young scientists are fictitious because I wanted no repercussion on them, although it is forty years later. Please forgive my condescending use of "Little" attached to their fictitious names—it is meant of a sign of familiarity. The words above are based on notes I had written in a hotel room in Hamburg, (West) Germany, where I had arranged to meet with climate scientists at the Max Planck Institute there—my second meeting with Klaus Hasselmann. I have edited those words only lightly, even though they were written when I was pretty tired. The "institute" or "laboratory" was the Main Geophysical Observatory, a very fine research institution where the two senior scientists referred to were Lev Gandin and Igor Karol, both of whom were very gracious and kind to me. The time in Novosibirsk was actually in a suburb called *Academe Gorodok* (Academic Town), a scientific town or village where scientists enjoyed many privileges compared to other folks. There are a few other anecdotes regarding that summer that I will tell in the rest of these memoirs.

The evening before the trip to Novosibirsk was eventful enough to mention. There was to be a going-away party for me at the apartment of one of the graduate students (I think his name was Serozhia; he is in the photo with a vest on under his jacket). His family was out of town, so the coast was clear.

Prof. Norbert Untersteiner, Atmospheric Sciences, University of Washington, who would become a close friend, was attending the meeting held in Leningrad before I was to fly to Novosibirsk. Norbert was Austrian by birth and about a decade older than I. He was single and always on the lookout for interesting women. Norbert was very friendly and curious about my visit to the USSR. Guena befriended him immediately and invited him to my going-away party. So off we went in a taxi on that last evening to the grad student's apartment. The apartment was on the opposite side of the Neva River and returning was not possible after 2 a.m. because of the drawbridges. Sitting next to me in the taxi was a beautiful girl with black hair and dark eyes, Natasha. According to Guena she was Cuban but had been living in the USSR for many years. Clearly, Guena was "fixing me up" with Natasha. I did not respond. I had been fighting off Soviet girls for seven weeks and was not going to give in now—not that I am such a prize, but because they all wanted a means of emigrating.

There had been an incident in the Buffet (a kind of canteen with small pastries, boiled eggs, and a tasty sour-milk drink at breakfast in the Hotel Europe, you never knew when this thing would be open for business— often it simply wasn't). A nice-looking Russian woman sat down at my little table and began to speak in English. She seemed fascinated with me. Telling me about how she owned a car, and how we might enjoy a swim together in Lake Ladoga, which was driving distance from Leningrad, I deferred, but naively gave her my room number. She kept calling, but I did not answer or return her calls. Finally, she got the message and stopped. I told Guena about it, and he said that I should avoid any contact with such people. I am sure there would have been trouble for me if I had responded—after all, I cannot swim!

Getting back to the party, it was wild and crazy. There were many people crowded into the (surprisingly nice) apartment, and at one point, according to Norbert, a corpse was taken down the stairs from a neighboring apartment. Eventually the police were called because of our noise— I was embarrassed that the party for me caused such a disturbance, but what do I know? Somehow Guena and other "adults" were able to shoo the cops away. The magic hour of 2 a.m. came and went. I was given a clean bed to sleep in. Norbert and Guena were still there. I was covered with bedbug bites despite the fine apartment. We had to rush back to Hotel Europe to get me ready to fly out for Moscow and on to Novosibirsk.

I stayed in Moscow overnight with my guide, Little Serozhia (different from the late-night party at Serozhia's). He was taking full advantage of his opportunity in the big city, and we attended a stage play. I could not understand what was going on but sat through it patiently. He told me it was quite modern. The next morning, we were off to Siberia. The experience in Novosibirsk was priceless.

I stayed in a kind of boarding house in Academic Town. My scientific guide was a man named Igor Kontorev. He spoke very good English and was clearly a high-level assistant to Dr. Guri Marchuk. Marchuk, an applied mathematician, was the head of the Siberian Branch of the Soviet Academy of Sciences (later he became head of the entire Soviet Academy of Sciences). Once set up, I was to give a seminar to a large audience. Kontorev interpreted my English into Russian for the audience. Like Gandin in Leningrad, he clearly understood the science as well as the language.

I made the lecture a bit more mathematical in hopes of impressing Marchuk, who was well regarded in the West. He was a first-class mathematician who worked on methods of interest to nuclear reactors (neutron diffusion) and also numerical models for weather and climate. When I gave my lecture, he sat in the front row. He was called out once for a phone call, but returned quickly. At the end of the talk, he stood up and reviewed the lecture for the audience, and added his opinion about it—I gathered this was a tradition at Academic Town. I asked Kontorev how Marchuk liked the talk, and he said he really did like it. I returned to my room, and soon Kontorev appeared, adding that Marchuk would like to have lunch with me. Of course, I was delighted.

Lunch was at a clubhouse for scientists. There were two visiting Bulgarian scientists who joined us. We sat on a porch and were served a full-course meal. Of course, there was vodka. Marchuk gave one toast after another for the Bulgarians and for me. His English was impeccable, even joking and kidding. As the meal went on, the vodka kept coming. The Bulgarians and I were getting pretty loopy. Marchuk continued his chatter. Suddenly, he looked at his watch and told us he had to leave for an important meeting. The Bulgarians and I fumbled through the rest of the meal and could not wait to get to our rooms and sleep. A large man, Marchuk was evidently unaffected by the alcohol. I guess this is a helpful trait for advancement in the Soviet System where vodka is so commonly served. But then, maybe his shot glass was being refilled with water.

I had some time for leisure in Academic Town. I was assigned a young and charming, red-haired interpreter. We went for walks in the birch woods nearby, and the only problem was mosquitos, large and aggressive. She took me by train to Novosibirsk where I met her boyfriend. He worked in the rare book section of the main library. He was generous with his time, showing me several Bibles, hand-printed by "Old Believers." These people had been resentful of the trend set by Peter the Great in the sixteenth century toward the Westernization of Russia. Many of them withdrew their families to remote villages to avoid persecution. Some of the Bibles were hand-copied centuries ago. That evening the interpreter took me to an opera, *Spartacus*. Of course, the opera was to praise the rebellious slave Spartacus as a precursor of the Bolshevik Revolution.

Georgii Golitsyn met me at the airport after my flight to Moscow, and he took me to his apartment, where his wife cooked a delicious meal. We had vodka with some kind of prairie-grass flavoring. The next day I was scheduled to visit the Science Officer at the American Embassy in Moscow. I was asked how I had been treated: did anyone try to enlist me for bad purposes, and so forth. I had been treated well at all times, and I was never approached for any nefarious purpose. I did not mention the woman at Hotel Europe. Of course, getting from the street into the embassy was interesting, with the Soviet guards making very sure that I had the appropriate credentials to enter. That was enough of the USSR for me in 1977.

I flew from Moscow to Hamburg, where I had arranged beforehand to visit Klaus Hasselmann, director of the Max Planck Institute. I had met Klaus a year or so before in Seattle at a workshop. At that time, we had a couple of nice chats about our mutual interest in stochastic climate models (models for which the rate of change of surface temperature is tickled by random numbers, imitating weather fluctuations). He had published a landmark paper in 1976 on this concept. I had come up with the same idea while in Boulder, but did not have enough experience at the time to make any progress on it. Later I did make such models and wrote a number of papers on the subject, extending the concept to global energy balance models with land/sea distribution.

I gave a seminar in Klaus's spacious office. I got the idea that such a talk was not common there. Or maybe he did not reserve a room, not knowing what I was prepared for, given that I was out of communications while in the USSR. His very fine group of climate scientists hauled in a screen and set up a projector and chairs in his spacious office. I proceeded to lecture to about a dozen younger scientists—the same talk I had given in Siberia. It was well received, and they put me up in a pension house for the night. This was new to me but satisfactory, and I was grateful. I walked around and found a pub where I drank a beer and had a (huge) plate of cheese and crackers. There was a standing shower cubical in the room with tubes that brought in fresh water and took out wastewater. I think the toilet was down the hall. It was all rather strange to me, but I was happy—and exhausted.

I met with Klaus and his associate Hans von Storch a number of times over the years, and we came up independently with the same method for detecting weak climate signals due to external forcing in the noisy climate system. Klaus sent one of his young star climate scientists, Gabi Hegerl, to College Station for the summer to work with my group. She ended up marrying my dear friend Tom Crowley—more about that later.

It was time to get back to teaching and my kids in St. Louis.

## Another USSR Trip: Tbilisi

In October 1979, I was once again invited to join a Working Group VIII group going to Tbilisi in Soviet Georgia (now the independent country of Georgia). The theme of this meeting was climate modeling. The leader of the American delegation was Eugene "Gene" Bierly, who at that time was in charge of atmospheric science research at NSF, and who funded my first government research grant in 1975–76 in St. Louis. I had many interactions with Gene, most importantly, his hiring of Tom Crowley to work as a two-year visiting program manager in paleoclimatology at NSF.

I already knew many of the Americans in the delegation, such as Isaac Held and Max Suarez, whose paper on energy balance models just preceded my first paper on the same subject. They both had PhDs from GFDL at Princeton, under the direction of Syukuro "Suki" Manabe, whom by now I knew well. Another old friend on that trip was Bob Chervin, a climate modeler from NCAR. Included was the legendary professor from MIT, Edward Lorenz (famed "father of chaos theory"), and his wife. As I understand it, he wanted to do some hiking in the Caucasus Mountains, such that he did not show up until the meeting was well underway. I had met Lorenz while I was at NCAR, where we struck up a friendship. He was well-heeled in mathematics, and he took a liking to my papers on energy balance models (EBMs), I think because of their elegance. In fact, he supervised the dissertation of Alan Robock, whom I knew later when he took a faculty position at Maryland and even later as distinguished professor at Rutgers; Alan was also on this trip. Alan's dissertation work came along just as I was simultaneously building an EBM in which noise forcing was inserted in the model to imitate the constant jiggling from weather (I mentioned this earlier in connection with Hasselmann). I once gave a seminar at Alan's request at Maryland titled "Noise in the Toys." My toy model solutions were analytical; Alan's were from numerical simulations. This reflects the cultural differences of our backgrounds (i.e., theoretical physics and meteorology), which was mostly a matter of aesthetics, each in its own way.

The meeting in Tbilisi was held at the Meteorological Center. One notable thing about this lab was that a guy named Joseph Dzhugashvili (later known as Josef Stalin) was an employee at this institution in 1899 when he was twenty-one (see the colorful book by S. S. Montefiori, *Young Stalin*). Actually, he collected data and made entries in notebooks for the meteorologists. He had been a student at the local seminary in Tiflis (now called Tbilisi). They raised the fees at the school, leaving him short, so he dropped out of the seminary. Apparently he was not paid very much as a "weatherman," but it left him with plenty of free time. It was just before this job that he had joined the Russian Social Democratic Labour Party. This was the beginning of his revolutionary activities and eventually his rise to lead the Soviet Union. Once while in a taxi in Tbilisi, I noted that the driver had a picture of Stalin prominently displayed on the sunshade in his car. I asked one of our hosts about this later and was told it was not unusual in Tbilisi. Many people there obviously still revered the old tyrant, who died in 1953.

Another rather odd thing to me in the city was that there were lots of foreign-made cars. I asked my friend Guena Menzhulin, who was himself of Georgian origin, and he said that there was a great deal of corruption in Georgia and lots of free enterprise operating in the black market (out of Moscow's sight). This meant that the time for waiting to buy a car was only a few months in Georgia, while in Leningrad it might be years. In fact, it seemed to me that the general standard of living was higher in Tbilisi, including clothing, shoes, etc. The central planners should look at this as a learning opportunity about economics.

A couple of social highlights on the trip are worth including. The first was a soccer game we attended. A large bowl-shaped stadium housed the event, and we had good seats at midfield. A large and lively crowd filled the stadium. Lots of police were periodically spaced on the track circling the field just outside the sidelines. I know little about soccer, but this occasion was a special treat. Ed Lorenz sat next to me, and we studied a program to try and decipher what was going on. It was written in Georgian and in Russian. Knowing the Russian alphabet, we determined that the Tbilisi team was playing the Moscow Dynamos. In fact, we could use the standings like a Rosetta Stone to see how the Georgian characters lined up against the Russian. I cannot remember who won the game.

The highlight of the social calendar was a trip to Tsinandali, a village in the Kakheti district of Georgia about a five-hour hot and dusty bus ride from Tbilisi. We, of course, had no idea where we were going and wondered whether anyone knew. We had not eaten a bite, and all were famished to the point of near fainting, but we survived. Finally, there was a beautiful palace-museum at travel's end. The well-known (but not to us) Chavchavadze family, headed by a prince, had owned the entire village. We toured the palace, where there were references to visits by Pushkin, Russia's most admired poet, remembered here by a portrait hanging on the wall. As I stood admiring another painting, my friend Georgii Golitsyn tapped me on the shoulder to point out that the portrait was of one of his ancestors.

The Golitsyns are very prominent in Russian history. There were many Prince Golitsyns who were confidants and advisors to the Tsars going back for centuries. I remember once in Moscow visiting Georgii, and above his desk was a huge oil painting of the mathematician Leonhard Euler, who was the world's leading mathematician in the late eighteenth century. Georgii showed me a big ring he wore that indicated his family heritage. I wondered how such open reference to the nobility could be tolerated in the Soviet Union in Brezhnev's time. Later I was told by a friend that it is not only tolerated but fashionable to have some royal blood. The reason seemed to be that the so-called "Decembrists," a group of "liberal" young nobles who tried to start an uprising in 1825. Czar Alexander I, who had been in power during the Napoleonic War, had died on December 1, 1825, and the uprising tried to take advantage of the change in leadership. Czar Nicolas I, who had just been installed, crushed the Decembrists and sent them to trial. Those who were not executed were sent to Kazan, Siberia, and other sites in the Far East essentially for life. Many wives followed their exiled husbands to take on lives well below that of their former stations in the highest levels of the aristocracy.

Georgii Golitsyn later became director of the Obukhov Institute for Atmospheric Physics in Moscow, and he has won many awards, including the A. A. Friedman (who was mentioned earlier in this chapter) Prize from the Russian Academy of Sciences. I recently read a translation of a memoir of Georgii's father, Sergei. Sergei was born at about the time of the Bolshevik revolution and lived a long and productive life under the Soviet Regime, but there were many close calls because he was officially a prince, as was Georgii later.

That evening our hosts took us to an outdoor dinner under trellises interlaced with grapevines. We were starving by the time signs of nourishment arrived. Behind us at the table were waiters with pitchers of wine made from these very grapes. Our self-control was weakened by our elevated appetites and the dusty deprivation we had suffered. We began to drink lots of wine. When it was finally time to eat, the wine consumption was starting to tell. Still, some were steady and brave enough to stand and make a toast.

By the way, after dinner, the Georgian tradition is to start with a toastmaster (called the *Tamada*), who tells funny stories and probably pokes a little fun at the host and some members of the party. I cannot remember who served in that role, but I heard Lev Gandin preside another time, and he was hilarious. After that, the *Tamada* calls upon others to give short toasts. There were obligatory apologies from Gene Bierly for the remarks at that time made by President Jimmy Carter, who on his high horse had scolded the Soviets about human rights. Our hosts were so hospitable and kind that we could not restrain ourselves from trying to explain. Isaac Held poignantly told of his parents (or was it grandparents?) trying to escape from the holocaust in the last days of WWII, and how the advancing Russian Army saved them.

I had delivered a very nice toast the night before when we were listening to a small ensemble of men singing in Georgian fashion. It was really beautiful. I must find a recording of that music. Anyway, the theme of my short speech concerned the lovely Georgian women whom we had seen. They did have a very characteristic facial look that was quite noble. This caused my friends to urge me to step up. I obliged by coming to my feet with wine glass in hand. As I stood for about sixty seconds with everyone waiting, it became obvious to everyone that I had forgotten any words I might have uttered. I sat down, and then the toasting moved on. After the feast was over, I had pretty much sobered up, but I found that many of my American friends were totally soused. Some very famous persons had to be carried to the bus. I have no idea when we arrived at our hotel sometime in the wee hours of the morning, for we all slept in transit. What a day!

Victor Frolkis and interpreter in Novgorod, 1976

Near Tblisii: Igor Karol (left),
Jerry (middle), Georgii Golitsyn
(right).

*Near Tblisii: Guenadi Menzhulin (left), Jerry (middle), and graduate student Seroshya (right).*

*Steve Schneider (left), Bob Cahalan (middle), and Jerry (right), ca. 1985.*

In Tbilisi: Ed Lorenz (left) and Milt Halem (right).

In Tbilisi (left to right): Isaac Held, Steve Espensen, Jerry, Max Suarez, and Bob Chervin.

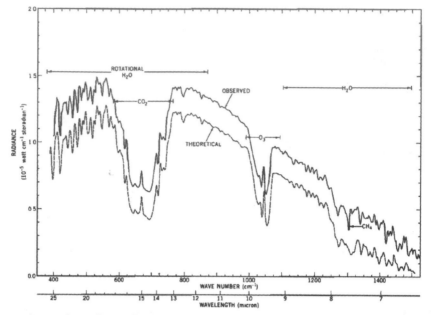

Radiance of upwelling infrared radiation over a site in the Gulf of Mexico where balloon-measured temperature data were available. One curve is the measurements from the IRIS spectrometer on board the NIMBUS 3 satellite. The curves are offset for clarity. One is observed, the other is calculated. All done on April 22, 1969. The figure comes from Conrath et al. 1970, Journal of Geophysical Research 75: 5831–57.

Beloved colleagues and coauthors while Jerry was at NASA/Goddard Space Flight Center (left to right): David Short, Thomas L. Bell, Robert Cahalan, and Jerry, 2018.

# CHAPTER 7

# CAMELOT

## *NASA/Goddard Laboratory for Atmospheres*

I took a job as senior research scientist at a NASA research laboratory. Making this move had an immeasurable influence on my work and my life thereafter. At that time, the environment at the Goddard Space Flight Center (GSFC) was uniquely suited for my background and skills. It was also a period at Goddard where a kind of "anything goes" freedom reigned for scientists. I also met and married the love of my life while there.

### Dave Atlas's Laboratory

It was a spring day in St. Louis when I received a call from David Atlas. I had met him briefly during my time at NCAR, where he was employed at that time. He told me he was now at NASA/GSFC, and that he was the new chief of the Goddard Laboratory for Atmospheres (GLAS). He intended to make it a rival of NCAR for excellent research, but with an emphasis on Earth observations from space, a new thrust of NASA's mission. He told me he had many people at NCAR and elsewhere tell him that I might be available and that I was highly qualified. I had already considered Miami and Illinois, and also a civil service job at NOAA in Boulder, for which I did have an unofficial offer from Joe Fletcher, who was a laboratory director there in charge of $CO_2$ and other greenhouse gas measurements from new sites around the world. I still had concerns about moving away from St. Louis because of the children.

Dave wondered if I might be interested in a branch head job for the Climate and Radiation Branch. I said I did not think I had enough government experience for that, and besides, I was not so interested in administration, but was a strong supporter of satellite data collection and the important application of it to the climate problem. In that very phone call, he offered me a research scientist position at the GS-15 level, which had a salary of $36,000 per year. I did not really know what that meant, but it was a lot of money considering that my salary at UMSL was about $25,000 per year, assuming I could hustle a summer salary from somewhere (I did have a small NSF grant, monitored by Gene Bierly, that covered me at least for the next couple of years.). I told Dave I would consider it, and he said he would call back in a day or two.

After a few days, he called back and asked if I had consulting income or something like that, because the raise in salary was large, and he might have trouble justifying it. I said nope, I am a poor man, and the size of the increase made me interested considering my family situation. He said OK, I will push the administration for it. He did, and they went for it. I accepted the deal, but I warned Dave that I was concerned about my family back in St. Louis. He was very persuasive, and his persistence convinced me. I took the offer in written form to my dean who had effectively mocked my earlier offer from Miami. He was astonished at the level of the offer, I suppose thinking I was still bluffing. Clearly, I just took it to him so I could poke him in the eye with my newfound market value. In my inflated imagination, I was a great titan who could lift the world if I wanted to (after all, I was friends with Atlas!). I wanted to. I took a leave of absence from UMSL, and all of my great colleagues there supported my decision and wished me Godspeed. I considered myself a new master of the universe.

Dave Atlas had been a full professor at the University of Chicago, then a senior scientist at NCAR before he took the job at GSFC. The academic world did not suit Dave or his career path because he wanted to do really big things, such as only an institution like NCAR or GSFC could pull off. Dave's specialty was radar, which he had learned during WWII. He was stationed in England during his service in the US Army Air Force. England led in the development of radar applications to spotting enemy aircraft. Many believe that England's lead in radar research and development was a key to winning the war.

Dave went on to win many awards for his work with radar, convective storm research, and later his great leadership. These accolades included winning the Rossby Award of the American Meteorology Society (AMS) and election to membership of the US Academy of Engineering. Dave told me that my letter supporting his nomination for the Rossby Award was crucial to his winning—he was prone to hyperbole when crediting his supporters.

In 1972 Wilhelm Nordberg and Robert Cooper from NASA/GSFC approached Dave just as he was suffering the consequences of his strong opinions regarding the (lack of) plausibility of suppressing hail by weather modification—he turned out to be right. Nordberg had achieved great success by observing phenomena from airborne sensors and then applying the knowledge gained to satellite instruments. His group had proved that they could retrieve quantitative information from satellite-borne sensors that were relevant to weather and climate. Originally, Nordberg was recruiting Dave to be a branch head for mesoscale meteorology (weather at the scale of a few kilometers).

Bob Cooper was the flamboyant director of GSFC, and he saw an opportunity to bring additional energy and innovation to a nascent Earth observations program at Goddard. Cooper wanted Dave to head up a new division to be called the Goddard Laboratory for Atmospheric Sciences, and he promised great flexibility in the form of an unlimited number of recruiting opportunities for outside scientists. Sadly, the key scientist and leader in the program, Bill Nordberg, was already dying from an inoperable brain tumor and had passed away before Dave arrived at GSFC. It was in this context that I received the phone call from Dave in 1978.

Dave hired outstanding scientists and essentially let them do whatever they pleased—an incredible thing to do in any kind of mission-oriented government laboratory. I dubbed it Camelot. AT&T's Bell Labs was an exemplar of that kind of research paradise, a setting where the transistor and many other great discoveries were made. It was similar to the atmosphere at NCAR in its early days. In later years, Bell Labs was shut down as Ma Bell was split up, both NCAR and GSFC were also soon to be plateauing in my estimation of their Camelot-scores. I could see it coming at GSFC and GLAS in particular during my last days at GSFC. Mind you, NCAR and GLAS continued to do great stuff, but the atmosphere was different. The university world was the only place left—but even there, it was not to be forever. Maybe research Camelots continue to crop up in the high-tech startup companies, or something similar. Of course, the enemy is accountability, the use of short-term metrics, visionless cost-benefit analysis, and the seduction by short-term measurable outcomes to make conclusions about long-term issues. Often these are dominated by a lack of vision high in the administrative pyramid.

## Earth-observing Satellites

The Nimbus series of satellites formed the second generation of Earth-observing spacecraft beginning in 1964—the first generation, the Tyros satellites, just demonstrated the feasibility of some of the basic observational concepts. The Nimbus satellites were more sophisticated, carrying a full suite of quantitative detectors, seeking information that could be used directly in potential applications such as weather forecasting and determining critical parameters about the global climate system. These satellites were called sun-synchronous orbiters. Their circular orbital trajectory passes nearly over the poles—these are called *polar orbiters*. The orbital plane of the satellite's trajectory is not quite perpendicular to the Earth's equatorial plane. Instead of exactly perpendicular, the angle is a little less than 90°. The orbital characteristics can be adjusted a few degrees as the satellite is set into its elliptical path such that it flies over a given spot on the surface at the same local time every day—for example, it might be adjusted to fly over at 7:30 a.m. and again twelve hours later, at 7:30 p.m., a phase favored by the US Airforce because observation of sky conditions at dawn is of particular interest for bombing missions.

For example, consider Nimbus 3, one of the early meteorological satellites, because its mission overlaps with issues I will be involved with later (TRMM and, quite separately, the greenhouse effect). Nimbus 3 was launched in 1970. It carried an infrared interferometer spectrometer that could report the spectrum of infrared radiation (the intensity of radiation that can be found in each of a *spectrum* of small bands of wavelength) coming up from the top of the atmosphere and going out to space. This measurement is like a fingerprint of how the outgoing energy is distributed across a range of different wavelengths. Understanding this distribution is a critical component of the energy balance of the planet. Making so precise a measurement was a singularly spectacular achievement. For the first time, it gathered information that could be used to see exactly how different molecules in the Earth's atmosphere absorbed and reemitted infrared radiation—all this in the real live atmosphere. In particular, it showed the unmistakable fingerprints of water vapor, ozone, methane, and $CO_2$ (the primary greenhouse gases) and how they are reducing the outgoing radiation from leaving the system. The intensity of the upwelling radiation in different narrow wavelength bands could even tell us about how high in the atmosphere the radiation was originated. These were fundamental observations about the climate system.

Simple radiative transfer computer models (c. 1969) gave excellent agreement with those observations, provided the temperatures and water vapor at all levels of the atmosphere are known beforehand. These were available simultaneously from some sites directly under the satellite's path as a result of balloon measurements. This was one of the first "ground truth" experiments for atmospheric measurements from space. The consistency between satellite measurements and calculations was a confirmation that they were doing something right. A great friend, Dr. Cuddapah Prabakhara, one of the coauthors of the Nimbus 3 work, sat just a few doors from my office for the eight years I worked at GSFC.

Another member of the series, Nimbus 5, launched in December 1972, carrying a group of instruments, some different from Nimbus 3 and 4. Of most interest for my work later in 1984 was the Electrically Scanning Microwave Radiometer (ESMR) for which Claire Parkinson, Joey Comiso, H. Jay Zwally and Tom Wilheit were the project scientists—both Claire and Jay became good friends of mine later. ESMR was a flat-faced device located on the bottom of the satellite. It looked down at the planet as it whizzed along in its sun-synchronous orbit, measuring the intensity of upwelling microwave emissions at a single narrow average across wavelengths. As the satellite travels above its ground track, this instrument's beam detector sweeps (not physically, which would require moving parts, but electronically—how ingenious is that?) back and forth across the ground track below, taking snapshots of the microwave intensity in roughly fifty-by-fifty-kilometer (thirty-by-thirty-mile) boxes. It covers not just snapshot boxes along the ground track but a swath about one thousand kilometers (six hundred miles) across. The microwave intensity is from an upwelling beam emerging from those small footprints below the satellite.

Each footprint's reading is a measure of the "apparent" temperature at the surface, but it strongly depends on the type of surface being viewed. The upwelling microwaves pass straight through clouds if it is not raining. Seawater has a very low emissivity (about 0.4) whereas ice at the surface or land has a high emissivity (close to 1.0, the value of a perfectly emitting surface). If the emissivity is unity, the apparent temperature is the actual temperature of the emitting surface. Flying over the ocean as you cross from open water to ice, the apparent temperature jumps by many tens of degrees Celsius (or Kelvin, to get Fahrenheit, multiply by 1.8). This contrast is a big signal (compared to the inevitable noise, only a degree or so Celsius) that even a pretty crude detector can read. So if only a fraction of the footprint is partially covered with ice, you get an apparent temperature in between. This provides a way of measuring the fraction of the area of the footprint covered by sea ice. NASA satellites used this discovery made by NASA research aircraft to infer sea ice coverage from space; the incomparable Bill Nordberg supervised much of this work. Records of sea ice coverage near both poles have been collected continuously since 1972. These began the stream of the astonishing sea-ice data—now routine—informing us about the loss of sea ice in the Arctic, year after year.

Another effect that was discovered in that same time period was that as you fly over seawater it looks "cold," but if there is a patch of the sea surface overlain by a rain shower, suddenly the footprint indicates a warm spot (the emissivity of the raindrops is close to unity and ESMR is measuring the actual temperature). In a seminal paper published in 1979, Tom Wilheit (much more on Tom later) showed that this kind of data provides a good estimate of the average rain rate in the footprint. This provides a way to estimate precipitation rates out over the oceans. This simple model doesn't work over land (no longer a big contrast in apparent temperatures over dry land and raindrops just above), but after many years, there have emerged methods (algorithms) to do that as well. In those wild and crazy days, a satellite vehicle (called a bus) and its package of instruments could be proposed, accepted, built, and flown in just a few years. Those were giddy times for scientists looking at new things from space for the very first time. Weather and climate conditions could be observed quantitatively from spacecraft. One had hardly the time to look at what came down from a mission before another one was being launched.

So even before Dave Atlas arrived, there was serious activity in Earth observations from space at GSFC—much of this pioneering effort was overlooked in the wider science community. Bill Nordberg (1930–76) had established a fine microwave group, and he employed creative aircraft experiments to learn what satellites might be able to do next. As already mentioned, Nordberg died just as Dave was arriving. Tom Wilheit, a young PhD in experimental physics from MIT, could not find a job in academia (reminiscent of my own case) and started as a National Research Council (NRC) postdoc in 1971 under the supervision of Bill Nordberg.

## Moving In

I moved to Maryland in September 1978. I found an apartment just outside the District of Columbia (DC) in Silver Spring, Maryland, about eight miles around the beltway from GSFC, which is located in Greenbelt, Maryland. I took a one-year lease on an apartment on the fourteenth floor of the Georgian Tower, just a short distance on Georgia Avenue from the beltway. I rented some furniture and settled in. There are a few stories worth telling about my Silver Spring experience.

Mike and Cheryl flew in for a visit the day after Christmas. The kids were busy and active as one might expect of a thirteen- and ten-year-old. They were sufficiently responsible that I gave them a key to the apartment. On about the second day, the three of us were on the elevator going up to the apartment when Cheryl was tossing her key up and catching it. Just as I was about to say don't . . . it dropped to the floor and bounced down the slit to Elevator Hell. Fortunately, I had my own key and we could still get into the flat. I made the call and sure enough, someone at the front desk knew how to recover their key and they did. This was all good, but perhaps an omen.

A few days after Christmas, I suddenly felt a pain on the left side of my lower abdomen. I knew that feeling from previous experience—kidney stone alert! I was figuring out what to do while doubled up on the floor in excruciating pain. Because of the holidays, everyone I knew in Maryland had scattered. No one in suburban Maryland had family there, because they all had migrated there from elsewhere—they were away for the holidays. Finally, I thought of Milt Halem, head of the modeling branch. I also knew that he lived in Silver Spring, not close by, but at least in the same suburb. I found his number in the book, screwed up the courage and called him. His tough, booming Brooklyn accent came through—everyone from New York sounds like they come from Brooklyn to me. I told him of the problem, and he came immediately, took me to the local hospital and took the kids home with him.

I was appropriately drugged, and overnight I unknowingly passed the stone and all was well. I have had ten episodes of kidney stones in my life—this was at the halfway point—each one different, some passed without bother or just vanished and some incredibly painful requiring a trip to the emergency room. I recovered one as it attempted to pass through a filter the doctor had given me. It looked all spiny like one of those burrs that you might pull out of your dog's fur—the kind that really pricks your finger and thumb. No wonder the damned things hurt so much.

A few years ago, I gave a talk at Princeton University, and the next morning I jumped on the train for the ninety-minute ride to Newark where I was to board my flight back to Texas. After about thirty minutes, I started to feel that old familiar pain. What am I to do now? It was not too terribly painful, but I figured the worst was yet to come. Somehow I seemed to have passed it, or it dissolved before I left the train. I was lucky that time.

This out-of-the-office experience brought Milt and me into a personal relationship that continues to this day—I always look for him at conferences. As branch head, Milt was a sort of "street fighter" at GLAS. All the other branches and their people were crap as far as he was concerned—I being an exception, because he always introduced me by saying I was from NCAR. (Who wants to be introducing a guy from a branch campus of the University of Missouri?) To me of course, UMSL had been a life-saver, but who am I to contradict Milt when he is introducing me to Jule Charney or Edward Lorenz (inhabitants of the pantheon at MIT)? In terms of quality and fame (and pedigree) of the members of his branch, of course, Milt was right about the superiority of his group. Especially, if you consider that from his view climate modeling was the only science going on at GLAS—instruments and the like were just engineering. Most people in GLAS outside his branch disliked (or were wary of) him, partly because he knew exactly how to pull strings to get what he wanted. For example, his branch put on a fabulous Christmas party every year, and we wondered how he found the money, the booze, and the rest. I suspected that his grateful contractors came up with everything for their employees, of course (his civil servant–scientists could always drop-in). It was just one of those tricks that made his troops love him and everyone else wonder why their own branch head couldn't do that. When they learned it might have been shady, they despised Milt all the more, but he knew exactly what he could get away with. I asked my branch head, Al Arking, about it, and he just smiled.

Over my time at GLAS, Milt brought in people like Charney and Lorenz, both of whom I met in his office while having a cocktail (forbidden in the Center, of course)—at the time Charney was wearing a rather ugly toupee due to his chemotherapy. Milt was also very creative in (legally) setting up appointments for affiliated scientists who came to GLAS (actually just to his branch), including Gerry Herman, assistant professor of meteorology from the University of Wisconsin; Michael Ghil, mathematician from Courant Institute at NYU; Mark Cane from Columbia University; Dave Randall from MIT; and many others. They would visit for months at a time, and Randall was actually hired by Milt, full time. I never quite figured out how those arrangements worked—but they did. Milt also brought Yale Mintz (recall Yale from my Tashkent trip) from his retirement at UCLA to work in his branch (as a contractor from the University of Maryland). Part of Milt's success in these endeavors was his vast knowledge of the great players in the meteorology and applied mathematics community (his own PhD was in applied math from NYU's Courant Institute). Most other branch heads (except Dave himself) were clueless about these people. They were deep into their instruments, orbits, and algorithms that brought in data from satellites—most were hardly into traditional meteorology (i.e., weather forecasting or climate modeling).

My second Silver Spring story came later in Georgian Tower. One night at about 2 or 3 a.m., I smelled smoke. I got up and walked around a bit, just as a voice came on the sound system telling everyone to get out of the building at once, but naturally "do not use the elevators." I wet a towel, put it around my head and started out. The hallway was full of smoke, but it was stratified such that the air was clear from about waist level down to the hallway floor, warm smoky air aloft. Everyone was crawling in the hallway, some crying or screaming. There were quite a few seniors in the building. Residents on the fourteenth floor made it on hands and knees to the staircase. Our crowd quickly descended about six floors. At that point, the staircase was full of smoke. All the warm-smoky stuff had just arrived at the upper floors in the stairwells. We all exited into the local hallway to try the other staircase, which turned out to be clear. We filed down to the street level and onto the sidewalk in pajamas, and it was pretty cold.

About thirty-five people were hospitalized from smoke inhalation, but there were no serious injuries. It turned out some fool fell asleep while smoking and passed out on a couch on one of the lower floors, and the smoke circulated throughout the building via the ventilation system—mostly in the hallways. Over the next few weeks, there were more warnings, all false alarms. Overly anxious residents probably triggered these after smelling someone's toast burning.

There was a parking garage in the basement of the Georgian Tower. This was very handy because in late February there came an incredible snowstorm that dumped about twenty-five inches of the fluffy stuff on the ground. Everything closed, including GSFC. I was marooned in the Georgian Tower, but at least my little red Chevette was dry in the basement garage. Even after a couple of days, it seemed too much for the little red car to venture out. At last, I was going stir crazy. I decided to call my new friend Laura Banks (Dave's secretary) just to chat. I had asked her out a couple of times, but we could never get our schedules to match. I had gotten to know Laura because I had been to see her many times in those first few months because of the government paperwork that was so confusing for new employees. Also, many clerical workers in GLAS were hopelessly inept (or lazy). I was writing manuscripts full of equations in longhand, and the branch secretary was elderly and simply could not handle the work. As the division secretary, Laura was the boss of all the aging secretaries in the division. So I got in the habit of going to Laura to get my manuscripts typed on her IBM electric. This was all before desktop computers.

Clearly, Laura felt pity for Jim Weinman and me because we were always getting crossways with the unfamiliar and often maddening bureaucracy. I slammed the phone down at the travel office because of the baffling paperwork and regulations regarding arrangements required for travel to a meeting or some such trivial thing that would have been no problem in a university—even UMSL where research-related travel was novel. On one occasion, Weinman actually resigned in the Travel Office. When notified of this scene, Laura had to get Dave to talk Jim out of it. There was a business assistant, Alberta, in the directorate office (one hierarchical level above GLAS). Alberta was one of those bureaucratic tyrants who could make life unbearable for the scientists, apparently just for the sake of asserting her administrative power. Once she called Laura, who managed to get along with her, demanding, probably with an evil smile: "Would you get your Dr. North to behave?" Little did Alberta know that I really was "your Dr. North," not just a guy in her division. Of course, when I had one of those rare visits to Alberta's office, she was all charm. But she could let a travel request sit in a pile on the corner of her desk until the last minute, driving us nuts: "Am I going on this trip or not?" In those days, routine travel in the United States was troublesome, but a foreign trip might require a signature downtown at NASA Headquarters with petty powerhouses letting us know who was really boss in the organization. At times, I thought maybe UMSL was not so bad. Of course, research at NASA was "science in a fishbowl"; the older a government institution the more the regulation—fear of scandal with financial consequences rules over creativity and innovation.

Returning to the great East Coast snowstorm, known as the "President's Day storm, of February 18–19, 1979," Laura was equally stranded by the event in her apartment in Greenbelt (near GSFC). Her car was sequestered under a two-foot layer of snow outside her apartment building. Her roommate Sara was stuck all this time at the group home for mentally challenged men where she worked as a supervisor. We talked for a few hours about the kidney stone, Milt, and my kids, who by then had returned to their mom. So after sufficient getting-to-know-you stuff on the phone, I mentioned that my car was dry in the basement. I suggested a treacherous test of the red car in traversing the eight miles of perimeter around the beltway. She said, that if she had known about the dry red car, I would have been on the beltway hours before. I packed some dry socks and my last package of peanut-butter crackers in case I might get stuck, saddled up and headed around the beltway, which was surprisingly navigable until I entered her parking area. She was waiting for me there with mittens waving, like those guys at the airport guiding an incoming plane into the gate.

Laura and I continued dating, but we kept it secret from our colleagues at GLAS. We went out to lunch a few times, dinners in downtown DC, and then to a Baltimore Orioles game at the old Memorial Stadium on the north side of the city. That was fun, but we were surprised at how wild the baseball crowd was. It happened to be "bucket night." Fans were served their beer in wire-handled metal buckets—each had a capacity of about a quart. Those buckets were swinging and slopping all over the stadium. Getting out of the parking lot later was more exciting than usual—passengers cheering, horns blowing, and general mayhem. It was the beginning of a mad love affair that continues to this day.

## Building and Diversifying GLAS

A large group of atmospheric modelers was brought from the Goddard Institute for Space Studies (GISS) in Manhattan to GLAS. GISS is located above the very restaurant in which Jerry Seinfeld, George, and Elaine gave us so many laughs in those days. I think relocating the group that became Milt's branch was intended to close GISS, but it failed—GISS is still running strong today. Over the years, there were many futile attempts to close GISS or at least move the whole operation to Maryland. The recently installed director, James "Jim" Hansen understood what a friend the *New York Times* could be in hard times. I think he single-handedly kept GISS in Manhattan.

Nevertheless, GSFC acquired a group that had developed an atmospheric general circulation model under the leadership of Richard Somerville (my friend from the 1976 summer at WHOI) and fashioned after Arakawa's digital methodology at UCLA. Milt Halem was made the branch head in charge of large GCMs in GLAS (at the Greenbelt, Maryland, site). There were two NASA GCMs, the other was at GISS.

Dave wanted a team of excellence and innovation in GLAS. When he arrived at GSFC, he had been granted lots of (Camelot-like) concessions from the director of GSFC, but he found a division that was spotty in terms of talent (and pedigree). James A. Weinman came to us from the University of Wisconsin where he was a professor of meteorology. Jim checked in at GSFC the very same day I did. Weinman was trained as an experimental nuclear physicist, also at Madison, but retreaded into a radiative transfer expert while going up the tenure-track ranks in meteorology under the tutelage of Verner Suomi, a world-renowned expert in satellite remote sensing. Jim was bored with academia and a real bohemian, divorced like me, he could not bear bureaucracy, also like myself. We found ourselves next-office neighbors in the Climate and Radiation Branch.

Our branch head was Al Arking., a PhD in theoretical physics from Cornell (he told me his advisor was Philip Morse). In government labs, a branch is sort of like a department in academia—usually a dozen or so government scientists accompanied by contractors that are mostly scientific computer programmers. Al was a radiative transfer expert, trained as an astrophysicist. Radiative transfer is the study of how radiation passes through the atmosphere—at some wavelengths, it is scattered; in some others, absorbed; in yet others, transparent.

Al was a brilliant and lovable eccentric. Rabbi Marc D. Angel memorialized Al a few years ago accurately and tenderly at his funeral.

> Al was born and raised in the Syrian Jewish community in Brooklyn. Although Al lived most of his life away from Brooklyn and away from his birth community, he was deeply rooted in the Syrian Jewish tradition. He loved the minhagim. He chanted the Torah portions regularly, in the Syrian Sephardic tradition. He conveyed to his children and grandchildren the uniqueness of his family tradition. Al's love of Judaism was at the essence of who he was as a human being. He studied Torah, he argued points of halakha (and I mean argue!), he sought truth. He and his wife Viviane were among the pioneers of the Orthodox synagogue of Potomac, and Al was a tireless activist in the congregation and in its Sephardic minyan. He was an outspoken advocate of the Jewish community and an ardent supporter of Israel.

Al could get into friendly but tangled arguments about science and government in much the way Rabbi Angel described in this passage about religion. Dave wanted Jim Weinman or me to replace Al as branch head—although beloved, Al was just too tiresome in meetings. The consensus seemed to agree that Al was brilliant and charming, but too eccentric in a bureaucratic environment. One personal peculiarity was that he procrastinated until the last minute to deliver stuff to his superiors and everyone else. I would see him in the office the afternoon before taking a trip with piles of work in front of him. When I dropped in for a moment, he would grin at me for a second, then continue grinding his way through, perhaps until midnight getting his arrears out. Weinman replaced Al. After less than a year, Weinman withdrew from the branch head job; it was agreed that he was even less suited, so Al was reinstated and it worked out. Al's eccentricity never got in my way. In fact, he gave me the freedom I wanted, pretty much unconditionally.

When I arrived, Al took me right in, gave me a nice office, and I had no duties except to do good research. He managed the funds I had at my disposal. At that time, I had no idea how money came to me or where it came from. Al simply said, "Don't worry about it." I had enough funding to hire a couple of contractors who worked for private companies but had offices at GSFC. I guessed Dave Atlas, the GLAS chief, obtained these funds from high up the food chain and dispensed it to the branch heads via negotiation or fiat, I have no idea. I never had to write a proposal in those days. Al said I shouldn't worry about it, and I didn't.

# CHAPTER 8

# ADVENTURES OF A GOVERNMENT SCIENTIST

Being a government scientist is different from being a professor. For instance, money for research flows differently. A government lab is always being watched for the rare scandal that might occur. Such things make traveling or dealing with contractors a different experience, depending on the branch of government.

## Adjustments

I worked as a civil servant research scientist at NASA Goddard Space Flight Center in Greenbelt, Maryland, from September 1978 to August 1986. My best and most lasting research was conducted in this period with the help of many admirable colleagues and support scientists. I arrived at an optimal time for a scientist with my education, experience, and peculiar set of talents. Over the course of the eight years I was there, my fit to the system was simply remarkable, but as time passed, my fit became less suitable as the organization tightened its range of scientific interests, and I was a poor fit as a government manager. However, I look back on this time with pride and fondness.

The climate science group grew by a few bodies, and I was asked to look after two newcomers, Tom Bell and Bob Cahalan, both retracing my own path as former high-energy theoretical physicists, coming through NCAR where they were postdocs. We hired a masters-level meteorologist from the University of Washington, David Short, who came highly recommended by his advisors Dennis Hartmann and Mike Wallace. These three men and I formed a group we dubbed the "twig," a branch of the branch (nerd humor). I was extraordinarily lucky because these three were creative, hardworking, and incredibly smart. I also had three contractors:

John Mengel, a PhD in astrophysics from Yale; Long Chiu, a PhD, trained under Ed Lorenz and Reggie Newell at MIT; and Fanthune Moeng, a PhD advised by Morton Wurtele in boundary-layer meteorology from UCLA. Fanthune's wife Chin-Ho Moeng, a PhD student of Arakawa, was a civil servant in Halem's branch. The contractors worked for a company called Applied Research Corporation (ARC), whose president was S. P. S. Anand, a PhD astrophysicist. He had been a postdoc at CalTech and an assistant professor in physics at the University of Toronto. He was to become a good friend of mine. I coauthored some of my best papers with these men and enjoyed a long friendship with Anand.

In the climate and radiation branch, my goal was to find problems that we could tackle with our peculiar sets of skills, honed from our training as theoretical physicists. I came up with the idea of mathematical statistics to be one of our themes. I do not think any of us had ever taken a course in statistics, but we were just arrogant enough to ignore that. Theoretical physicist and Nobel Laureate Richard Feynman once said, "Science is the belief in the ignorance of the experts." Meteorologists that were turning their attention toward climate science were always trying to study functions—temperature, pressure, water vapor—of location and elevation on the sphere. For example, the numerical value of a field of surface temperature depends upon where you look on the global sphere. Another is the field of pressure, which is also a function of position, and both evolve in time. The more I read about these studies, especially their statistics, the more I became interested.

## Our "Home Run" Paper in Statistics

I got to know the head of the department of meteorology at the University of Maryland at College Park, Ferdinand "Ferd" Baer. When one of the professors retired, it meant that the graduate-level turbulence course was without a teacher. He asked me if I wanted to teach it, and I agreed. I taught the turbulence course for two years, learning the subject as I taught. When Prof. Anandu Vernekar took his sabbatical leave, it meant that his statistics course needed a teacher. I was paid about $1000 for teaching each of these graduate-level courses, but it was worth it for my brushing up on (OK, learning) these subjects. I had never taken a course in these subjects, but I was eager to learn them.

During the semester I taught the stats course, I drove over to the campus with one of the masters-level contractors from Halem's branch, Dave Gutzler, who had been a colleague of Dave Short's at the University of Washington. He sat in on my statistics course and that caused me to work especially hard to do a good job. In fact, I learned enough statistics to teach, but I also delved much deeper beyond the course needs into the subject. Confession: I am selfish—I teach mostly to learn. My physics education allowed me to learn these subjects. But I must say that learning statistics was a great and pleasurable experience for me. I got so into it that some people thought I was a trained statistician. Gutzler went on to MIT to work for his PhD under Ed Lorenz and later spent his career as a professor at the University of New Mexico.

Meteorological researchers did not seem to me to have a very deep understanding of statistics. I became familiar with an area called random fields—some of this was inspired by my Leningrad friend, Lev Gandin— one of the reasons I am so taken by my visits to the USSR. This kind of statistical model turns out to be a very convenient way of thinking about the fields that are of interest in climate science. A random number is one where its value is determined by the outcome of an experiment, such as tossing a pair of dice. Each toss leads to a different value (the sum of the numbers on the individual die). Each toss is referred to as a *realization* of the process. There is a probability distribution of outcomes: the value of seven occurs often (because there are so many combinations that add to seven, snake eyes and boxcars are rare [only one way out of thirty-six combinations to get each]).

In the 1940s and 1950s, statistics offered some promise in weather forecasting because there were available techniques from other sciences that could be borrowed. Such forecasting techniques are pretty much the "only game in town" in the field of economics (so-called econometrics). In the late 1950s, dynamical equations of meteorology (the laws of physics) as implemented and solved numerically on the digital computer sprang into the picture as the competitor to statistical methods. By the 1970s, the dynamical/numerical approach was running away with that show, and people forgot that statistics can still be important, especially when it comes to climate applications where we might want to know whether a change in climate might be pure chance or deterministic (noise and signal). Another application in meteorology involved testing to see whether seeding of clouds could enhance rainfall (generally, good statistical tests show that it is problematic at best). The youngsters coming up in meteorology programs were so enthralled by new-fangled dynamical methods, which were so much more elegant because they were physics-based and beautifully formulated, that they began to ignore the all-important random aspects of their science. This was the prevailing *paradigm*.

I became very troubled by the low quality of papers that employed statistical methods. In mock retaliation, I coined a (fantasy) parlor game: "Given any paper in meteorology that purports to use statistics, contestants have ten minutes to find the error." The most common error is sampling error. Serious scientists, especially in sociology and psychology, learned long ago how to use statistics properly in their research involving finite samples of subjects.

In my group at GSFC, the ideas, usually started with me, were shaped and focused through give and take with my colleagues and assistants. These players faithfully generated the software to implement the ideas; then we pushed our creations to concrete results that went into the scientific literature. Our discussions were exciting and productive. I wrote all the manuscripts with pencil and paper. A secretary then typed them. I do not suppose anyone could ask for more. Everything was slow compared to now (no cell phones, no email, laptop computers). The senior civil service members (especially Bell, Cahalan, Lau, and Short) of the group cooked up their own projects as well as those generated by me. They were brilliant.

One of the first projects that I supervised had to do with random fields. I suggested that Bell, Cahalan, Moeng, and I look at a problem important in comparing random fields from one another, for example, a field of data versus a field generated from a climate model simulation. Climate scientists frequently study the geopotential height field—a measure of the horizontal and vertical distribution of atmospheric mass. For example, it is useful to examine the level (in meters) of the height where half the mass is below and half above—it is called the 500-millibar-height field. Fluctuations in this surface carry valuable information—such as spatial and temporal correlations—about storms and other features. Predicting its future shape is important in weather forecasting. It is possible to reduce these fluctuations into a series of "mode patterns" known as empirical orthogonal functions (EOFs). These patterns were similar to the sinusoidal shapes of a violin string vibrating for individual pure tones, except here they are in two or three dimensions—in the case of two dimensions, think of a drum head; for three dimensions, a cavity of air such as in the volume of a cello or a horn.

We decided to check out the sampling errors in estimating the shapes of the EOF patterns. We cooked up an artificially generated random field (we knew how to do this) that had the same statistical behavior of the geopotential-height field and tested how sensitive the EOF patterns were to the size of the sample (number of realizations of the measured or simulated field). This turned out to be a "home run" paper for us. As soon as I saw the result, I knew we had something important. I said, "Stop now, I'm writing this up for publication." I had the paper written in longhand in just a couple of weeks. It involved mathematical techniques I had learned in my quantum mechanics course way back. My coauthors, Cahalan and Bell, were right with me because they had taken the same physics courses years before. Moeng was our expert programmer on the project. Later, Moeng became a successful meteorologist at NOAA.

That paper, titled "Sampling Errors in the Estimation of Empirical Orthogonal Functions," was submitted and published quickly by *Monthly Weather Review* (North et al. 1982). That journal was a venerable old periodical in meteorology, but popular at that time in the emerging field of climate science, where there were not specific journals available. Years later, I was checking the statistics of citations, and I found that our paper was the second most-cited paper in the history of that journal. At the time of this writing, our paper had over 3031 citations as reported on the Google Scholar Website—over 2042 citations on Web of Science. The technique is often referred to as "North's Rule of Thumb." I was the lead author, but my coauthors deserve much credit for this work.

Yale Mintz, a man of many maxims and rules, quoted me an old rule of his regarding the inclusion and ordering of authors of a paper. There are three aspects: 1) who had the experience to know the problem was important and suggested it; 2) who actually did the work, the heavy lifting, the experiment or mathematical solution; 3) who actually wrote or led the writing effort of the paper. Often this is easy, especially if one person did two or more of these components.

## Two-Dimensional Climate Model of the Planet Earth

One of my favorite projects was the creation of the two-dimensional energy balance model, which we had built mostly in order to study paleoclimatological problems. I had been thinking about this model since my days at NCAR. I was too inept as a programmer to implement it on the big computer. I knew the exact numerical method to apply, but there was no hope that I could ever get such a program to run error-free. I had approached several potential collaborators, but for one reason or another, no one wanted to undertake the job. When I got to GLAS, I found that huge pool of talented scientific programmers. Fortunately, the first guy that I started to work with was John Mengel, who worked as a contractor for ARC. Anand assigned Mengel to me as a scientific programmer. John had a PhD from Yale in astrophysics and was familiar with the techniques of numerical modeling of problems somewhat like mine. I also got Short involved because of his excellent intuition about climate and statistics. What a team that was! We met almost every day discussing progress and the next moves. Tuning a model—adjusting some of the coefficients (a cynic would say "fudge factors")—is a bit like trying to find a radio station on a shortwave receiver, kind of an iterative procedure. In our problem, it involves turning several knobs in the fine-tuning. Short's intuition was critical.

From the outset, it was not obvious that such a model would work at all in terms of describing the seasonal cycle of Earth's surface temperature field, but I would not let go of it. My intuition said it would work. Once running, the model reproduced the data for the distribution of the present seasonal cycle of surface temperatures extremely well. We demonstrated the dominant influence of the land-sea distribution on the planet in determining the mean seasonal temperature distribution and the geographical dependence of the winter-summer swing or amplitude of the seasonal cycle. The results were astounding, demonstrating that the response to the solar heating going through the annual cycle was essentially linear. We showed that the sinusoidal amplitude and phase of the annual cycle, as well as that of the semiannual cycle, were in near-perfect agreement with the data. We even conjectured in the paper that we were on the path to explaining the growth of ice sheets in the Milankovitch Cycles (forty years later, I still believe it). The paper was titled "Simple Energy Balance Model Resolving the Seasons and the Continents—Application to the Astronomical Theory of the Ice Ages" and published in the *Journal of Geophysical Research* (North et al. 1983). This is the paper that caught John Imbrie's eye, and it sent me to London and the Geological Society of London symposium to present the paper. Prof. Richard Peltier and William Hyde (the latter then a student at the University of Toronto, but later to become my postdoc at NASA/GSFC) went on to use our model to include glacier models for the ice sheets.

There were other papers in this same genre from our team. I won't go into them here, but the level of creativity and productivity pouring out of our group was remarkable. I went on in later years to publish many scientific papers in climate science, but these were the times I look back on with the greatest pleasure of my professional life.

## The Sun and Climate

In May 1981, while a civil servant at GSFC, the leaders of Working Group VIII, in cooperation between the United States and the USSR on pollution and climate change, called upon me once more to join a delegation to the USSR. This trip was designed to examine the evidence for the variations of the Sun's effects on climate change. John "Jack" A. Eddy, a friend I had met while I was at NCAR as a senior fellow, led the delegation. Alan Hecht of NSF co-led in his role as a government official and for his expertise as a paleoclimatologist. Alan was a grant monitor at NSF for climate dynamics—more recently he has served as sustainability advisor to the assistant administrator of the US Environmental Protection Agency.

It was a rocky time between the two superpowers, and there were hijinks going on that made our trip a bit nerve-wracking. For example, at the last minute, our language interpreter was denied his visa to enter the USSR. In fact, all visas were held at NASA Headquarters (HQ) until the last minute. I had to go to HQ on the day of departure to pick up my visa, and I was given visas for several others to be delivered at the airport in New York—a ridiculously close shave at best. Finally, the problems were solved at breakneck speed, and we were all airborne for Moscow.

At the airport in Moscow, we had a tough time getting through the security checks. Alan Hecht had a popular paperback spy novel, and the Soviet official would not let him bring it in. I had heard from previous visits that such books were put under the desk and sold later on the black market. Being a smart aleck, I demanded that if they keep the book they tear out the last fifty pages where the story winds up. I insisted that they destroy them in our presence to make the book worthless on the black market. After some haggling, they let the book go through. But next, after going through security, I was detained in a room with large windows viewing the waiting area where everyone sat waiting to learn of my fate. Could Siberia be next? Again as my natural smart-aleck genius came into play, I spread myself out and read the *Wall Street Journal*.

After about an hour's wait, they passed me through. Jack Eddy was told that the date on my visa was off by a day. I was coming into the country one day ahead of the date on my visa! (My conspiracy theory was squashed— fortunately, I was not to spend the rest of my life like the Decembrists!) They told him it might be that I would have a problem at the hotel, where all visas and passports are collected and kept until checkout. Actually, I did not have a problem at the hotel. They seemed to understand my predica- ment. Afterward, my colleagues declared me "Hero of the Delegation." What was I thinking? One thing I had learned that my colleagues did not know was that simple but polite resistance in the Soviet Union often actu- ally works in these circumstances with petty immigration functionaries. Often, they do not want trouble any more than you do, especially when it might be embarrassing in the area of international matters. Avoidance of embarrassment in front of foreigners is important in the USSR.

Our leader Jack Eddy was a solar physicist in the High Altitude Obser- vatory at NCAR when I was there on sabbatical, and we had met. Jack was famous for a landmark paper he published in *Science* in 1976 titled "The Maunder Minimum." Changes in the Sun's brightness should lead to consequences to Earth's climate. Centuries of astronomical research tells us that the Sun is a typical star as found in studies of stars and their classi- fications. The sun is a *variable star* in that it undergoes a mild cyclic behav- ior at a period of roughly eleven years. The evidence for us today is that there are dark spots on the Sun's surface that appear, grow, move slowly toward the Sun's equator and disappear. The Sun rotates with a period of about twenty-four days. The lifetime of the spots is long enough for them to disappear as the Sun rotates and then reappear later on the way around. Galileo was probably the first to study them as a scientific phenomenon with his new telescope in 1610. Chinese scholars recorded them without telescopes as early as 800 BC.

The most interesting period for sunspots is a time span of fifty-two years between 1647 and 1699 when there were no detectable sunspots at all. Jack's paper revives the history of this discovery by a husband-and-wife team at Cambridge, the Maunders. The Maunders studied the records and made note of the fifty-two-year hiatus and published on it in the early twentieth century. Jack's paper made the term "Maunder Minimum" the standard way of referring to this peculiar period. The Maunder Minimum coincided approximately with a time when the Earth's temperature was declining into a period called the Little Ice Age. Jack drew attention to this connection, but he was cautious in his discussion not to infer that they were related in the sense of cause and effect. Although adequate paleoclimate data were limited in quality and sample at the time of Jack's paper, there docs appear to be enough evidence of a global cooling that has stuck to this day.

The possibility that global climate change was due to the Sun's variability excited the solar science community. Many papers showed up announcing new evidence of the connection. Sadly, many of the papers purporting to show new evidence were knocked down, usually because of poor statistical analyses. As elsewhere, in the Soviet Union there was plenty of support for this attractive idea. There was clear competition between the followers of the solar origin and the greenhouse gas origin of climate change. Budyko supported the idea of increasing $CO_2$ being the culprit, while his successor as director of the MGO, Borisenkov, appeared to favor the Sun's brightening. The two were never seen by us in the same room together, and it was clear to the Americans that this was a blood feud. Of course, we stayed out of it.

Along with Jack, the leader of the delegation, the rest of the American delegation was filled with highly respectable and yet very personable men. Let me enumerate some of them. First, there was the great carbon dating expert Minze Stuiver born in the Netherlands, but at that time professor in the Quaternary Research Center (QRC), the University of Washington. A few years later I visited Minze's lab while I was in Seattle on other business. It was a remarkable site well below ground to minimize the noise from cosmic rays. His paper with colleagues (Stuiver, et al. 1998) is the definitive study on correcting the results of radiocarbon dating using tree ring data going back ten thousand years.

Another member of our delegation was Murray Mitchell, a

paleoclimatologist at NOAA for thirty-five years. Murray was highly respected and neutral on the causes of climate change. Although as a scientist he was open-minded, he was concerned enough to warn about the possibility of man's role in the rising temperatures.

I visited Murray from time to time in his NOAA office during my time at GSFC. We continued to share an interest in how the Sun might affect climate. He did coauthor a paper in 1979 with Stockton and Meko, citing evidence that droughts in the US Southwest might exhibit an exceptional variance at a period of twenty-two years. This could be the result of a solar influence. Twenty-two is double the eleven-year period, and it is relevant because studies have shown that those magnetic field loops flip in polarity at the end of the eleven-year cycle, making the actual period twenty-two years. I mention it because it was an important paper and because Charles "Chuck" Stockton, a dendroclimatologist at the University of Arizona was a member of our delegation as well.

Peter Foukal, a solar physicist with a PhD from McGill and some time as a research scientist at Harvard, was another distinguished member who became a good friend of mine. He helped connect me with the solar community. Peter is president of a private company that does solar research. He also wrote a technical book about the Sun that I read after this trip. We have kept in contact and later wrote a short perspective paper in *Science* together (Foukal et al. 2004).

Finally, I will mention another member, Marvin Geller, whom we encountered in chapter seven, and he will appear again in the next chapter. Marvin was an expert on the Earth's outer atmosphere and the stratosphere, both of which are affected by the solar cycle through the variation of the Sun's luminosity and through the solar wind.

The meeting was in Vilnius, already an important city in the Middle Ages, the capital of Lithuania, then a Soviet Republic and part of the USSR, now an independent country. Vilnius (sometimes referred to as Vilna) has a nasty history of persecution of Jews. It may be that this is the region from which my childhood friend Benjamin Joffe had emigrated just before WWI. Violent pogroms in Jewish villages were common in those times. Vilnius also figured badly in the Holocaust. Marvin took a special interest and wanted to visit the local synagogue. I remember that the leader on their side, Dr. Mikiliounis saw to it that Marvin received a tour of the old Jewish section (Ghetto) of Vilnius and its old synagogue. It was an emotional experience for Marvin, probably the only time I ever saw him so taken. Ninety-nine percent of the time he is either all-business or all-jolly, mostly the latter.

I should mention that we took a side trip to the beautiful city of Kaunus and, of course, to the site of the tree ring lab that was on the Baltic Coast. The countryside is bright green on the grassy plain toward the shore. The beach is famous for deposits of amber that can be found by tourists. These deposits were laid down about 34 million years ago, perhaps as sea level was receding. They are especially noteworthy for the presence of insects killed instantly and embedded in the clear amber. Such finds are of interest to paleontologists.

We learned about the Soviet interest in the Sun through the formal technical presentations, which tended to focus on tree ring studies. As always, the best part for me was bonding with and learning from the American scientists through walks together or at meals. Our general impression was that the tree ring laboratory and the science happening there was pretty weak. The lab director was very enthusiastic in showing us the eleven-year cycle in the tree ring data, but we remained skeptical. The enthusiasm for this signal to appear in the data is common throughout the community of tree ring people as well as the solar people. Usually the length of records is not long enough for a reliable statistical test of significance.

## An Adventure in Leningrad

The most interesting thing that happened on this trip was after the meeting in Vilnius was over. Following is a short personal report of my trip to Leningrad after Vilnius to visit scientists at the MGO. I reproduce it here with only minor editing. It was sketched in a hotel room in Stockholm after the meeting in 1981.

It was Sunday, June 7, 1981. I had left the Soviet Union and stopped over in Stockholm on my way home from a symposium held in Vilnius, Lithuania. I left the other Americans in the delegation last Tuesday and went on to Leningrad to visit some scientific institutes there and renew some old acquaintances in that memorable city. It was my fourth trip to the Soviet Union so that most of the peculiarities of the Soviet system and the Russian mind were becoming familiar to me. But this little story provided me with some new insights.

In my previous visits to the USSR, one scientist especially impressed me—perhaps because he seemed to like my work. He was an older man, fifty-six at the time, but keen as though he were a young scientist absorbing new knowledge like a sponge. Lev and I became fast friends. His English was excellent, and he even served as my translator for the three hour lectures I presented at the MGO in 1977 (I later wrote them up, and they were translated into Russian and published as a monograph or "brochure"). We had lunch together every single day in the MGO cafeteria (lots of beets, cucumbers, and cabbage soup), and afterward we would sometimes walk and have a smoke. Lev smoked those peculiar Russian cigarettes called Papirosi—a cardboard tube with a short cigarette at the hot end—very strong smoke, almost narcotic. He had a way of crimping the end in the Russian style and lighting it with a wooden match, then carrying the Papirosi between his first and second fingers with the match between his second and third fingers. Lev named me "Severov" since "Sever" means "North" in Russian. Once he searched the town over for a package of Papirosi with the brand name "Sever" so that I could have this namesake as a souvenir. The closest Russian first name for me was Gerasim, so he named me Gerasim Severov.

Lev loved to joke, and his English was so good that he was able to make plays on words with ease. He constantly told anecdotes, occasionally with a political flavor, but I never detected in him any anti-Soviet sentiments. He hated bureaucracy as do I; we frequently commiserated on the subject. He often asked about problems in the United States such as gun control, racism, and drugs. Since I freely admitted that our system has faults and difficulties also, I think he began to trust me. I did not proselytize our way of life, nor did he, his. My admiration for him grew.

Lev was a brilliant man; he has a broad range of interests and had attained a level of excellence in each. During the war, he was so talented in music that he was an orchestra conductor and violinist in the Soviet army. He was also an excellent bridge player, and I have heard that he is unbeatable at table tennis.

It was during the war that Lev met Budyko, and Budyko persuaded him to join the MGO—also, they were notorious bridge partners. I have always thought that Budyko's promotion of Lev and other Jews to high positions in the MGO were the reasons (among others) for Budyko's dismissal after eighteen years as the director. Lev stayed on at MGO as Budyko moved on to found another group at the SHI, where I also have many friends.

When I was in Leningrad four years before, I had several companions relevant to this story. One was Vitya, a young Jewish graduate student whom Lev had recommended to be my guide. In this way, Vitya could improve his English, and I could improve my Russian. I liked Vitya, but found him to be a bit lazy, although very smart, and I was constantly annoyed at having to wait for him, so I learned to get around on my own. Once I had dinner at Vitya's home. His father, a medical researcher, was out of town and his mother, an English teacher, cooked an excellent meal and spoke with me at length. She was anti-American, defending every aspect of Soviet society. I suppose it was one of the few times I felt uncomfortable because I usually made it a practice to stay clear of sensitive subjects except to ask genuinely sincere questions about how various things work. I never opened such a conversation. Vitya liked me very much and wanted to spend as much time with me as possible, but we never discussed politics. I found it peculiar that when I left Leningrad to spend a week visiting Novosibirsk, Vitya was not permitted to accompany me—instead another grad student, a Russian, was sent along.

Another character from my past that figures in this story is Irene, a translator at the MGO. I first met her in Tashkent; she interpreted my talk there. We had a kind of flirtation going in Tashkent but let's leave that.

When I arrived in Leningrad for my extended stay, Irene was there and willing to accompany me to the opera and ballet. We had a lovely time, but this time I forbade any romantic involvement because I had a commitment back home; she agreed. We saw each other frequently as friends. Irene is a very intelligent, complicated, Russian woman. She is more cultured in literature and music than I, which made it a special pleasure to attend cultural events with her. Once she took me to a movie based upon the story "The Negro and Peter the Great" by Pushkin. We sat in the back of a nearly empty theater; she translated line-for-line for me. I like her very much. She does not like Jews for some reason and was rather frank about it. Still, in most matters, I feel she is an honest and sincere person. She was rather direct in conversation, which may have caused her to have few friends.

Igor Karol is a scientist at the MGO of about the same age and rank as Lev. His English is good (although Lev corrects him mercilessly), and his research interests are also close to mine. Igor is very meticulous and precise about everything from science to shoelaces. Russians told me that Igor is not Russian but German by descent and this, according to them, explains his meticulous nature. He and Lev were involved in editing and translating into Russian a monograph I wrote during my stay in Leningrad. Igor had a way of trying to help me that always ended up insulting me. I could never believe the suggestions he made because I thought they were nonsense. Igor usually had lunch with Lev and me.

Novgorod is an ancient city near Leningrad (about two hours away by car); I wanted to visit it during my summer visit to Leningrad in 1977, but my visa did not include travel to Novgorod. Once, during a coffee break, I learned that a young Jewish graduate student named Sasha was going to Novgorod on the bus to run some business errands for MGO; I suggested that I go along with him, see the town, and return with him that evening—no problem with the visa if you do not stay the night. It seemed a satisfactory solution to all of us, and I was very happy. I mentioned it at lunch that day and within an hour I had a knock on my door from a member of the International Office informing me that since my visa did not include Novgorod, there could be no visit. Igor had gone straight to the authorities. Not only was Sasha Jewish and known for his dissident antics, but he worked for that rival institute headed by Budyko. I went into a minor rage telling them it was not illegal, and I knew better. Lev tried to comfort me saying that Igor was only trying to follow the rules and that after all Sasha had been in trouble several times before this. The next day, I was informed that in a few days I would visit Novgorod and be taken by a private driver accompanied by Vitya (trusted) and a pretty translator. We went and had a lovely day—apparently, it was legal after all. This incident was so Russian, at least in those days.

I attended another symposium in Tbilisi, Soviet Georgia, in 1979. Irene was not there but Igor was. We renewed old acquaintances, and I inquired about Lev. Igor told me that Lev was well but too busy to attend the symposium. I wrote a warm letter to Lev with many of my recent preprints and other manuscripts for his reading and comments. In particular, I asked if he knew of any Soviet work on the subject which might have predated mine. Lev soon wrote to say that, as I suspected, Obukhov proved my theorem many years ago in his dissertation under Kolmogorov. Lev provided me with references and a few useful remarks. This news came first informally in a New Year's greeting, then about a month later in an official letter through channels.

A few years later, Igor sent me a copy of the printed monograph I wrote. It was a nice surprise, because I had given up on the whole project thinking it would never get through the Soviet review process. I heard that one of Budyko's rivals blocked the publication out of pure malice toward Budyko, but out of the blue, the reprints arrived one day with a note from Igor, translated into beautiful English by Irene. It was about this time that I learned of my upcoming trip to Vilnius, and I wrote to both Lev and Igor thanking them for the long ordeal necessary to get the monograph published. I also told them that I was about to be married and that I hoped to visit them on the way back from Vilnius. I soon got a letter from Igor and Irene saying they looked forward to my visit. I also received a letter from Lev informing me that he had retired from the MGO on his sixtieth birthday and had applied for an immigration visa. I was shocked; I had no idea he intended any such move. He was secure with a very good pension and enjoyed status as a scientist. Before retirement, his salary was several times the average; he owned a condominium, and his wife did not have to work. By Soviet standards, Lev was well off. In his letter, he suggested that I visit him at his apartment as a scientist; he included his address and phone number.

I had often wondered why Lev was not a member of the Soviet Academy of Sciences. He has made scientific contributions that were clearly as good as some people who are at least a Corresponding Member (a slightly lower category in the Academy). Budyko is a Corresponding Member; stories are whispered about how his rivals prevented his becoming a Full Member. I was told by a scientist in Novosibirsk that Lev was never admitted, not because he was Jewish, but because he was so irreverent; he might be admitted and make fun of them the next day in public. I just cannot accept that explanation; I know some other Jewish scientists (such as Yudin) at the MGO who are equally qualified but not as vocal as Lev. Like me, Lev was an emotional man, and when he began feeling the pressures after Budyko's removal, he might well have uttered some things not altogether in his own best interest.

In Vilnius, I met Igor and Irene. Over several conversations with Igor, I never mentioned Lev; Igor never did either. On about the eighth day, I mentioned Lev's name in connection with some scientific question and added that I heard he retired. Igor said "yes" but did not elaborate. Lev had become a nonexistent person.

I came to Leningrad by overnight train with Irene and other Soviet colleagues in the same coach. There was lots of beer drinking on the train, and naturally, I joined in. I was in a sleeper car in which there were two double bunk beds on each side of the compartment. I was assured by Guena Menzhulin, Budyko's right-hand man and my good friend who was on the trip as well, that I would get one of the lower-level beds. This was imperative because there was no ladder up to the second level. Guena explained that lots of Polacks rode these trains, and they stole the ladders (Lithuania is on the Polish border). So, after lots of beer consumption and singing, it came time to go to bed. Oops, my lower berth had been nabbed by a lady with a baby. This was a problem because I was too wobbly to get into the upper berth. Someone had to get up and give me a boost. I survived, but this was a tricky one.

When we arrived in Leningrad, I found that my hosts, the MGO, had arranged only a two-hour visit with Budyko's group whereas I was expecting to spend the entire day there. It was a busy three days in Leningrad with small conference meetings with scientific groups. On Tuesday night, I had dinner with Irene and then walked for forty-five minutes back to my hotel along the Fontanka Canal. It was midnight when I got back to my room; the sun was just on the horizon.

The next morning, I was astonished when the phone in my room rang. I picked it up and the voice said "Jerry?!" It was my wife, Laura. Since I had stayed on after the meeting in Vilnius, the others came home a day or two earlier. Alan Hecht had called her to fill her in on our trip, and he told her about my visa having the wrong dates, and my being named "Hero of the Delegation." By the time he hung up, she was petrified. We had only been married a few months, and although she knew of my bizarre travels, now I was her husband, and she did not cherish the idea of relocating to Siberia or any other such place. I reassured her that all was well and that I was not violating any laws.

On Wednesday morning, I tried several times to call Lev from the hotel but there was no answer; finally, someone answered, but we were cut off. That night, I went to the opera with Irene. I decided my Russian was not good enough to place the call so I turned to Irene for help. She managed to trace him down at a friend's house. Lev invited both of us to come to his apartment on Friday night. Irene said she thought it might be too risky for her because she had already been reprimanded for spending too much time with the Americans. I looked forward to Friday night with a kind of chilling anticipation, although there was no danger for me because as far as I know, there is no law forbidding my visit. Lev told me four years before that he was not allowed to have me visit the apartment then because the Party Bureau at the MGO would not grant permission. Also, I did not visit Igor's home, as he explained, for the same reason. All visits to private homes were done on the sly, if at all (recall that I did visit Georgii Golitsyn's apartment in Moscow in 1977). Lev was to meet me in my hotel lobby at 7 p.m.

Friday came rather quickly in my busy schedule, and no one suggested any cultural activity for me that evening so I did not have to make any excuses. It was odd that I was not able to talk with my student friend and guide, Vitya, all week. Once I saw him in the courtyard, and I thought maybe he saw me. With Igor tugging at me, I went into the yard, grabbed Vitya, and shook his hand saying hello. He seemed a bit flustered, and when Igor tugged at me again because we were late for a meeting, I looked around and Vitya was gone. I saw him one other time in a small group meeting, but he was whisked away with the others that worked there. Vitya was Jewish, and so our time was limited. In fact, I was rarely left alone for fear I was a spy (I guess).

I was in the lobby promptly at 7 p.m. I was absolutely shocked to find Irene standing there. She said she had changed her mind on Lev's invitation and had called him earlier in the day. Irene was visibly shaken as Lev walked through the door with his son, Nik. Lev and I embraced, introductions were made, and we were off to catch a taxi. In the taxi, Lev and Irene told me that they saw Borisenkov, director of the MGO (successor to Buyko) in the lobby; she was worried because she might have been seen with me. Irene seemed to think that quite by accident Borisenkov was there to greet a friend staying in the hotel. I think this is true since he is hardly an inconspicuous tail. I suspected we were being watched but not by him. Frankly, under these circumstances, I like them to have an idea of what I am up to so they can follow if they want to and see that all is aboveboard. Of course, words to this effect are never spoken, but when calls are made from hotel rooms or conversations are held in the rooms, it is assumed that you are being recorded so no need to discuss it.

We had a very pleasant dinner at Lev's with his two sons who speak English and his Russian wife who was learning. We made toasts and had many warm exchanges of thoughts, gifts, and scientific papers. Lev and I had only a very few moments alone together because it was not advisable to discuss any sensitive matters (such as his emigration plans) with Irene present. In those very brief intervals, I was able to piece together that a condemnation proceeding had taken place at the MGO because Lev had applied for an emigration visa. Some of his closest friends had been key in testifying against their old colleague of twenty-five years.

That night, I did not sleep much. The next morning, I had to get up to take a ride out to the town of Pushkin for a little sightseeing before catching the plane to Stockholm. I had seen Lev, but I feared for him. He assured me that all would be well. Later I learned from him that his brother had emigrated, and he felt that was too much for the authorities. It was also argued that he had been privy to classified information at MGO. After I returned to the United States, I was told by other émigrés that Lev had become a non-person. All of his scientific papers in the library had his name as author removed, so that no one could refer to his work as his. In researching for this book, I found that he was restored to his "person-hood" after the Soviet Union dissolved. Also, Budyko had become a Full Member of the Russian Academy of Sciences.

Lev was finally granted an exit visa, and he was allowed to come to America, where he was quickly hired at NOAA. He conducted research on quality control for the data assimilation process that is necessary for preparing a numerical weather forecast. I met him a couple of times in Washington and even had dinner at his apartment with his wife. After the dissolution of the Soviet Union, there must have been some changes in the MGO. As I perused the internet for a picture of Lev for this book, I noticed that he is now duly recognized, along with Budyko, as one of the most important scientists in the history of the MGO.

## Peat Bogs and Pottery

My last trip to the USSR occurred in 1989. The trip was focused on paleo-climatology. The leader of the delegation was Stephen Porter, director of the QRC (Quaternary is a geological term referring to the last four million years before present) at the University of Washington. Porter was a few years older than I and possessed a genial, handsome, leadership look. He was an expert on glaciers in North America. We had met a few years earlier at the QRC, where he had invited me to present a lecture at one of their colloquiums. The main host on the Soviet side for this Working Group VIII meeting was Andrei Velichko, director of the Institute for Geography in Moscow. Members of the group included Gorgii Golitsyn, by then director of the Institute for Atmospheric Physics in Moscow, and Budyko, SHI, Leningrad. The other Americans included Dr. Erik Sundquist of NOAA, a geochemist and expert on the carbon cycle; Paul Collinveau (1930–2016), an ecologist from Ohio State University—Paul was famous for a book he wrote about the maximum size that animals could attain based on physics (Colinvaux 1978)—and finally, Prof. Herbert Wright, a senior geology professor at the University of Minnesota, known for his studies of how the last large ice sheet in North America withered away and how the meltwater might have caused a thousand years of cold temperatures in the North Atlantic, the so-called Younger Dryas period, about 11,500 years ago.

The purpose of this trip was the sharing of scientific information via the usual lectures given by all with the usual overhead projectors and uncomfortable headphones. We were to focus on paleoclimatology covering the recent past few thousand years. My main purpose was to schmooze with old friends and get to know some new ones, picking up science as I went. Each of these provided a great learning exercise for me.

One suggestion I made was the need for pollen data and its analysis along a latitude circle from the western end of Russia to the Pacific Coast. Such data indicating climate change over the last ten thousand years could provide a test for climate models because the Milankovitch forcing indicates that summers in the interior of Asia should be much warmer than at present. Stephen Schneider and I coauthored a paper in a volume containing lectures given at the 1984 NATO conference in Grenoble, France. The signal should be strong since it appears in the EBMs as well as in the GCMs. I do not know if my suggestion was ever carried out, although John Mitchell (probably unaware of our work) at the Hadley Centre in the United Kingdom published something along these lines later. While at the Grenoble (Laboratoire Interdisciplinaire de Physique), we had a tour of the laboratory and the ice core archives.

We took an unforgettable field trip from Moscow about 250 km (150 miles) north of Moscow in a van. Our destination was near the city of Rostov. Rostov is on the shore of Lake Nero and one of the "Golden Ring" of cities with spectacular onion-domed churches from the late Middle Ages. An interpreter and an anthropologist also accompanied us on the trip. We were taken first to an unforgettable Russian village where one of the Russians had an elderly aunt that lived in a small cottage there. We were invited in for tea. Her cottage was old but spotlessly clean and truly authentic. In the corner were religious icons, and there was a large fireplace that extended into the large room. It was "plastered" over such that on cold nights one might sleep atop the fireplace. The area behind the fireplace was a room for livestock, which at one time slept indoors and helped keep the place warm. She was a lovely lady and most gracious to receive us.

Then we got into a motorboat and buzzed several miles across the lake. We disembarked from the boat to find ourselves in a flat plain of peat devoid of any plant life. Peat is a spongy layer of material that is burnable as a fuel. Over time it might convert into a form of soft coal, a process requiring pressure from overlaying material. It is dry, especially below the surface. It looks and feels like the shavings of a pencil-sharpener. There are tiny pockets of (usually dry) air embedded in peat. It is common across the subpolar lands of Alaska, Canada, Scotland, and Russia. About a half a mile away, we could see bulldozers scooping up the peat for use as a fuel or in some cases as a mulch in gardens. You can purchase bags of peat at stores specializing in gardening.

We hiked away from the open field, which had been cleared, and back into a wooded area, finally reaching a small opening in the forest. The excited anthropologist began jumping all over the place. He went to an exposed wall of peat sediments, scratched a little, and ancient reindeer teeth fell out of the banked peat. In a few more minutes out came some shards of pottery that he claimed were produced by lakeside dwellers about five thousand years ago. He laid out some of these on an opened newspaper on the ground. I was quick to see and point out that the headline of the newspaper had something to do with Mikhail Gorbachev's "perestroika," cameras snapped all around. Not every reader is old enough to remember that this was Secretary Gorbachev's program of openness or transparency. It was the beginning of the dissolution of the Soviet Union. What an experience!

On the bus ride back to Moscow, we stopped to tour an ancient cathedral in Pereslavl Zalessky (probably), which is a semicircle of towns with Moscow roughly at the center, famous for their magnificent Russian Orthodox churches. The artwork in these structures is stunning.

## More USSR Encounters

In between these visits to the USSR, my Soviet science friends would come to the United States for one reason or another, and Laura and I would welcome them to our home for dinner and drinks. Among these were Georgii Golitsyn, Igor Karol, and Vladimir Alexandrov from Moscow University. They all wanted to know what I was working on, and I gave them reprints of my papers. Alexandrov was a physicist who had gone into climate modeling, and he went over my EOF paper with me, line by line. I think he had a photographic memory. I always suspected that he was KGB, and perhaps some of the above were as well, but I never had any reservations about sharing my work on climate science with anyone. After all, my work was openly published and had little or nothing to do with national security.

A CIA agent called upon me at my home after some of my visits to the Soviet Union. The agent was collecting general information. I was happy to bring them up to date on the status of climate science in the Soviet Union. Patches of excellence in their science were evident, usually in the areas of statistics and applied mathematics. In my opinion, the imprint of the famous academic ancestor, Kolmogorov, was often clear (the leaders of the atmospheric and oceanic sciences institutes were direct academic descendants of his). The CIA agents would always ask me whether I was ever approached even mildly to become an agent for the USSR. My answer was always negative. Of course, like every scientist visiting the USSR in those days, the Soviets regarded us as potential American spies. But there was never any open attempt to "turn me" or to hint at giving them information that they should not have.

Recently I was on a cruise with family up to Halifax and found a book, *Transit Point Moscow* (1986), in a used book store. It is a memoir of Gerald Amster, an American in the USSR in 1976, and coauthored with the professional writer Bernard Asbell. This was at the time of my first visit to the

USSR. Amster was arrested in Moscow for drug possession and had to spend some time in a Soviet labor camp (prison). His discussion of life at that time and place in the USSR matches mine precisely as to the culture and other peculiar (to us, of course) ways of their people. I recommend it if you can find it. Books by the excellent travel writers, Colin Thubron and Ian Frazier on their experiences (although post USSR) also provide glimpses into some of the differences between our and their cultures.

After the Soviet Union dissolved (the time of Boris Yeltsin's reign as president of the Russian Federation), times were horrible for Soviet scientists. Many were reduced from super status to penury. Those who could do so emigrated. I know of many who came to this country, and I know many others who have gone to other countries. The wholesale emigration of scientists from the USSR is reminiscent of the same phenomenon from Germany and Austria in the late 1930s.

I loved my experiences with Russia and Russians. I tried during my 1977 summer to learn the language, but I was never very good at it. I can usually scratch my way through scientific text with a dictionary and possibly my old (wonderful) text by Potapova. It helps if there are equations and figures with captions, where there are many cognates with English.

I wanted to know what it was like to live in a communist country. I never felt that I wanted to live in a communist society, nor do I have the slightest communist sympathies, but I thought that the Russian experiment was one of the great events in world history. I read many books on the Russian Revolution, including biographies of Marx (e.g., by Isaiah Berlin), Trotsky (especially the three volumes by Isaac Deutscher), Lenin (by Robert Service and others) and Stalin (a good one by Stephen Kotkin [2015], but also several by Simon Sebag Montefiori). I wondered what would drive intelligent and creative people to do such things as they did—the answer turned out to be terror. One has to go back to the history of Tsarist Russia to see what caused people to take to such a screwy idea, resulting in the deaths of all those innocent persons. Part of it is deep in the Russian character itself and their fear of chaos. They prefer a dictator, as we see today. The other part was the sickening way the Russian nobility dealt with its people (see *The Romanovs* by Montefiori [2017, also 2005, 2008], Richard Pipes [1974], and many others.) Who could have developed the system of exile to Siberia? It was there long before communism, and weak forms exist even today. And the Russian character is well described in the great nineteenth- and twentieth-century Russian novels that I have read and reread (Gladkov 1994; Rasputin 1995). If Russia changes in these particulars, it will not be easy or quick. I submit to the theories of George Kennan, whose memoirs and historical works I have read along with his biography by John Lewis Gaddis (2012).

Also, I enjoyed and benefited from meeting and spending private as well as formal time with the great American scientists that were along with me on the Working Group VIII trips. Aside from the Americans I met on the trips, I worked hard to understand the plight of distinguished as well as less than distinguished Soviet people I met and engaged with. I felt for the cleaning woman in Hotel Europe in Leningrad who wanted to buy my blue jeans for her son. You had to wonder how and why people will betray their longstanding friends for fear of their own safety or status. But being in someone else's shoes in another place and time is not so simple.

## Marriage and Beginning a New Family

I was not just at my giddy career peak, but I was also in love. In March 1981, Laura Banks and I were married. It had been a fairy-tale-whirlwind romance. It was not likely to work, because of our age difference (fifteen years). I brooded that the gap was too great. Though she never brought up having a family, I knew Laura would be a great mother, and I knew that I would have to understand that I would be having more children, and I was forty-two years old. Was I nuts? In hindsight, perhaps, but not in this matter.

We were poor. First, we bought a townhouse in Seabrook, Maryland, a long but reasonable walk to my office at GSFC. My older kids were in private high schools in St. Louis and already considering college. I was paying child support and a good portion of their tuitions (Jane's mother also chipped in). Before we moved to Texas in 1986, Michael was at Swarthmore and Cheryl was at Bard College, both pretty expensive, but now in hindsight, perfect for them. It was a time of huge inflation and our mortgage rate was 11 percent. My next son, Adam, was born in 1983, and Joel was born in 1985. We stayed home and watched movies on TV. We never ate out. We bought a movie VCR player that included lots of rental coupons. We watched old movies for entertainment. I remember fondly the garden we had set up in raised beds in our (postage stamp) back yard. As a wedding present, Laura's sister Martha gave us a wonderful tan-and-white cocker spaniel. We named her Jitney, after the name for buses we encountered in Nassau on our honeymoon.

Poor as we were, we did sometimes entertain with cocktail parties at our townhouse because it was so close to GSFC. I was on the Scientific Colloquium Committee, and we hosted quite a few visiting lecturers in those days (e.g., Dick Lindzen, Mike Wallace, and Jim Holton). It was an honor that I was asked to give one of these SC lectures in 2016, usually presented by "bigwigs." I also gave one of the MANIAC Lectures in September 2018 (available on YouTube).

At these events, our little townhouse would be packed. Once, we had such a crowd over, and Richard Lindzen was our guest of honor. Dick had not yet become known as a notorious climate change skeptic, but he was already a contentious fellow (although one-on-one, quite charming) even before climate skepticism overtook him. Nevertheless, I found him fascinating, and he was so smart and quick that I greatly admired him (I still do). Dick and I squared off in the living room and commenced to debate whether smoking cigarettes was a health hazard. I had quit smoking a couple of years before, but Dick was a heavy smoker. We went around and around debating the subject. While Dick's position was pathetic, he was so skilled at argument that he beat me hands down. Decades later, he and I debated at the Houston Forum, he taking the skeptical side of the climate change argument. Again, he took the wrong side that could not possibly be defended, while I had the safe side with all the evidence behind me. He won again. There are some people that I place so far above me in the pantheon of scientists, in whose presence I simply fold up.

So my time at GSFC in Maryland was, as I've said earlier, a "Camelot" or magical time for me. Not only did my scientific output and creativity soar, but my home life, though austere, was equally magical. How lucky I was to have met the right life-mate for me, now happily married for thirty-nine years, and we had two older wonderful children to enjoy as well. However, I was beginning to perceive signs of displeasure with my membership in the civil servitude cropping up.

*Rostov, a city near the peat bog. (left to right) Herb Wright, Eric Sunquist, Jerry, Paul Colinveau, and Steve Porter.*

*Cottage near Rostov, USSR, where the US delegation had tea and a warm welcome before going to the peat bog.*

160

Pottery found in the peat bog near Rostov USSR. The newspaper underneath reads (in English) "Peristroyka" as declared by Secretary Gorbachev.

(Left to right) Jerry, Laura, Georgii Golitsyn in Seabrook, Maryland.

Jerry speaking at a meeting in Vilnius, Lithuania. US members present include Jack
Eddy, Peter Foucal, Alan Hecht, Marvin Geller, Minze Stuiver, and Arthur Few.

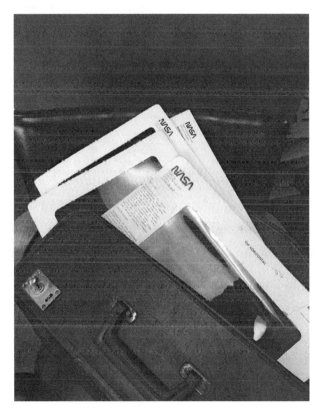

Typical overhead
slides with cardboard
borders used in TRMM
presentations shown in
a heavy carrying case.

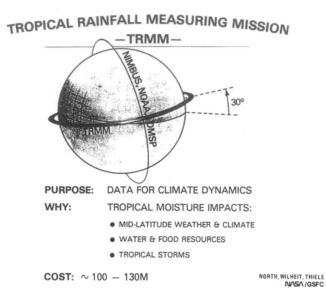

*Typical content of an overhead slide during one of the many "sales pitches" for TRMM.*

*(Left to right) Gene Rasmussen, Joanne Simpson, Jerry, Marvin Geller, Tom Wilheit, and Fugono's daughter.*

# CHAPTER 9

# TRAVELING SALESMAN FOR TROPICAL RAINFALL MEASURING MISSION

Certainly one of my greatest thrills at Goddard was my involvement in the Tropical Rainfall Measuring Mission (TRMM). In 1984 Otto Thiele, Tom Wilheit, and I proposed that NASA and a foreign partner build and launch a satellite that could measure precipitation in the tropics.

## Otto Thiele

One of my dearest friends at GSFC was Otto Thiele, eleven years my elder. For the first time in my writing this memoir, my eyes tear up as I remember Otto and reread his obituary. Otto was the administrative manager of the Climate and Radiation Branch, of which I was a member. Otto took care of all the material and financial details of the branch, allowing branch head Al Arking to roam "care-freely." Otto understood the Climate and Radiation Branch and the NASA bureaucracy from top to bottom. He was a true Texan, born in Austin, and a devoted Republican. His ancestors were farmers near Fredericksburg, located seventy-eight miles west of Austin, in the beautiful, but hardscrabble Hill Country. This is Lyndon Johnson country (see especially volume 1 of Robert Caro's multivolume biography of LBJ and the book by the great nature writer John Graves, *Hardscrabble*). It is one of many immigrant towns in Central Texas, each of which carries a different lineage, such as German, Polish, and Czech along with other Central and Eastern European ethnic groups. For example, two Czech towns, Snook and Caldwell, one Protestant and one Catholic, respectively, are near College Station. These counties usually have a town near its geographic center that serves the farms and related businesses. Often in these communities, some of the older people still speak their ancestral German or other native languages, but of course this tradition is withering. Some storefronts still show central European names on their signage. The quaint town of Fredericksburg is perhaps the most famous of these communities. Otto's ancestors were hardened by the unpredictable flood/drought swings, the rocky limestone soil, and the harassment from hostile Comanche raiders. They were a tough people, honest, straightforward, and decent. They believed in hard work and community efforts in bad times: it is pure Texas.

Growing up, Otto often spent summers at his ancestral farm near Fredericksburg. He joined the Merchant Marines in 1944, at age seventeen. He was a commissioned officer in the US Navy through the Korean War. Afterward, he returned to Austin and enrolled at the University of Texas, where he received a bachelor's degree in meteorology in 1957, the year of Sputnik's launch. He took a job at NASA and became the director of NASA's Apollo tracking ship, USNS Vanguard, stationed in Cape Canaveral, Florida. Otto retired from NASA twelve years after I left GSFC, but we kept in touch, and in retirement he often worked as a contractor for his friends at NASA HQ, organizing panels to review and grade grant proposals. I served on some of these during my time at Texas A&M. During these meetings, Otto usually had me over to his family home for dinner.

Because I was divorced and ignorant of government when I came to GSFC, Otto took me under his wing. One of the more memorable experiences was his persuading me to join his "duckpin" bowling team. After Laura and I were married, I continued to play. We were so poor in those days, but I used a large fraction of my weekly allowance to register for bowling and buy one beer.

Otto was the reason I jumped into the space program in contrast to my "pure" science focus. He told me about a new opportunity that had come up in the spring of 1984. There were meetings going on at GSFC about the possibility of a "new start" for a low-cost (end to end, less than $100 million) Earth observing project. The Nimbus series was finished, and its findings were turned over to NOAA; NASA was for building new technology, not routine data collecting, at least in those days. This new start opportunity was the first in a few years, because of altered priorities (away from environmental issues) in the Reagan administration. In the space program, the priorities were space shuttles, the Space Station, and other manned space flight programs—shiny Buck Rogers stuff.

I frequently had to remind myself that NASA is basically an engineering organization. Costs escalate because the satellite has to be elegant and filled with sexy instruments—sometimes sexier than is required—and not always about science. It is about engineering—sometimes science is just an excuse. It must be conceded that the space program from its beginning was about national defense. Astronomy is nice, especially to establish world leadership. But never forget that putting these objects in orbit was proof that we could deliver a missile's load in the backyard of any country we selected.

Sometimes we conduct good scientific research when no one is looking (a cynic might say this is true in universities as well). We had to expect that getting a new start in any discipline would face fierce competition due to tight budgets. It was no longer the "anything goes" days of the younger NASA eras. Also, all missions would have to be more expensive because NASA could not afford mistakes—like science in a baker's window. There were also distributional issues that decision-makers had to address. Every subdiscipline of science and geographical region had to have its due portion of the pie.

Scientists at GLAS were beginning to cobble together proposals for the competition. Otto argued that the tropics were very important in the climate problem because of a lack of data and the gigantic expense of gathering information from field experiments. A satellite that could somehow measure the components of the energy cycle over the entire tropics might be attractive. I started reading papers on the tropics and especially a review article (Houze and Betts 1982).

## Tom Wilheit

Otto told me that there was a guy named Tom Wilheit that I should get to know. Tom was the head of the Microwave Branch in GLAS at that time. Tom had been thinking about this subject for a long time. Recall that he grew up (after PhD) in the Nordberg group and was an expert in microwave radiometry. He even had an old radiometer "on the shelf" that was a backup from the Nimbus series of polar orbiting satellites. It was ready to go, but it had not found a flight opportunity.

I met with Tom, and we found that we had some overlap in that we were both PhD physicists, which was itself enough to form a bond because of all the standard education and scientific language we had in common. After our PhDs, we had both jumped into the midst of a terrible job market. That he was an experimentalist and I a theorist also had some complementary advantage. Tom was an expert on rainfall retrieval by microwave radiometry, having written the most important paper on the subject (Wilheit et al. 1977). He agreed with Otto that the tropics might be a great application for this kind of technology.

Tom was the leading expert on precipitation, and I had some expertise in the climate business—again some complementarity. Looking back over the years, I think he was the best of his generation on this subject. However, he was busy looking after his branch (mind-boggling bureaucracy), and he feared he did not have the time to take on the leadership role at this point. In my case, I had finished a whole pile of papers as already described, and I had no obligation that would hold me back. It was clear that Tom was a fighter in the rainfall community, and he feared that he had made quite a few enemies over the years. As I found later, his characterization of the foe issue was true, a common problem among high-stakes bidders in the NASA family. I knew no one in the rain or tropical communities, but I knew lots of meteorologists, and I had virtually no enemies. Tom had already floated a short proposal involving the energetics of the tropical part of the world and the role of precipitation. By energetics I mean measuring all of the energy components such as infrared radiative cooling to space, absorbed solar energy, and finally the latent heating in tropical clouds, which can be derived from precipitation data. The idea is to estimate flows of energy in the tropics, and by adding up the budgets for the individual components, checking to see if there is a balance. This is a common approach in large-scale circulation studies in meteorological and climate research. If possible, this might be done from a single spacecraft with the appropriate sensors on board to operate over a period of years.

Tom laid out the components of a mission that would fit this role. We could design an orbit that was tropical as opposed to the usual polar orbiters; that is, its orbital plane would be inclined from the Earth's equatorial plane by about thirty degrees (later it turned out to be thirty-five degrees). By covering only half the Earth's area, we would enhance our sampling rate at a particular location significantly. An important selling point was that there are essentially no precipitation data over the tropical oceans except for a few short-lived and very expensive field experiments. Later I once estimated that if we took all the ships in the world and fitted them with radars, there would not be enough ships to do the job of estimating precipitation at a good resolution over the world's tropical oceans. Another consideration in our favor was that the bulk of the surfaces in the tropics' surfaces are covered by ocean. It so happens that Tom's very elegant retrieval method works best over the open ocean, making the bulk of our retrievals more accurate—exactly where there are no rain gauges.

The most intense El Niño ever recorded at the Galapagos Islands occurred in the Northern Hemisphere during the winter of 1982–83 with serious environment (especially biological) consequences. In the middle latitudes, heavy precipitation and damaging winds in California led to some houses falling into the Pacific. The weather patterns in the mid-latitudes of both Americas were affected by weird weather. For the first time during an El Niño, there existed satellite imagery of clouds over the Pacific. These images showed unusually tall towers of cumulus clouds near the equatorial dateline.

The weather research community took notice. It presented a great opportunity for the observation of this phenomenon in all of its manifestations. The early general circulation models quickly leapt into service to help sort out the connection between the remote equatorial Pacific and the middle latitudes (mainly North America). The 1982–83 El Niño was such a dramatic event that not only the research community but the general public were talking about it through newspaper and TV coverage. Aside from the newspaper reports, footage from aircraft of heavy flooding was all over TV.

Early general circulation model studies were also hinting that this heat release aloft could excite long stationary waves in the atmosphere's pressure field that slowly propagates away from the equator and eastward into North America. The result is a distortion of the normal weather patterns across the United States. Our proposed satellite could measure that rainfall near the dateline and therefore provide important data for sharpening forecasts in the middle latitudes. Even in normal times without El Niño or its twin, La Niña, this cycle of energy release is the most energetic component in the global atmospheric system.

Given these circumstances, our message adjusted itself slightly from a mission to measure rainfall rate as accurately as possible at a site toward a mission that might contribute more to climatology or climate science. Our product would be month-long, area averages of 500 by 500 km² (500 km = 300 miles). We envisioned maps of monthly averaged rainfall rates over a grid of the areas stretching from latitudes ±30°. Tom and I noted a political consideration: NOAA was the main agency for weather applications based on satellite observations, and NASA was the lead agency for stratospheric studies (the ozone hole was at last nearly understood) and for climate. The El Niño range could be considered on the time and space scales of climate. Such data were needed to improve the large *climate* models that were being perfected for El Niño-type, as well as broader climate change, studies—although we would not stress the latter, due to political reasons. We could put forth the need for improved large climate models as the reason that NASA should control the mission. I was beginning to think that both Tom and I had a feeling for political leverage. Much later, after TRMM was flying, we found that one of the most important applications would be the observation of hurricanes, but during the community support gathering (sales) period, we dared not mention this because if TRMM were seen as a hurricane satellite, it might be argued that NASA was the wrong agency for such a mission, because NOAA was the agency in charge of such weather research, not NASA. We figured that if it became a NOAA mission, it was doomed—sadly, NOAA has no clout for new missions: she was just not sexy enough. TRMM had to be a climate science mission out of NASA.

The mission could also be pretty cheap, with its main instrument already available. We would need a visible and infrared sounder for simultaneous cloud observation, but this was a fairly cheap addition to the bundle. If our chances of winning became serious, we could also think about having a ground-focused, cross-track scanning radar. In terms of retrievals of precipitation rate, the microwave radiometer would be more accurate than the radar, but the radar could give us information on the vertical distribution of raindrops and perhaps shed light on the vertical distribution of the hydrometeors (millimeter-and-larger ice particles and raindrops). In principle, this information could tell us about the vertical distribution of latent heat release that is so important because of its influence on distant midlatitude storm patterns due to atmospheric wave motions.

There were four challenges that Tom had already identified:

1. Sampling errors, and he had a rough back-of-the-envelope guess of approximately ten to 15 percent (which turned out much later to be spot-on). This refers to the number of times the satellite passes over a given grid box in a month.

2. The diurnal (day-night) cycle, which was completely unknown over the world ocean. If its statistics were strange, it might lead us to biased estimates depending on the visit schedule of the spacecraft (always recording at the same time of day leads to bias!).

3. The beam-filling problem, an error in the retrieval algorithms that depends on the linearity of the rain rate versus apparent (microwave) temperature (unfortunately, it is not linear). This error is sensitive to the size of the footprint and the wavelength of the microwave beam as well as the correlation statistics of the rain field.

4. The ground truth problem, which was tough because it would be difficult and expensive. Arguments among hydrologists indicated that ground-based estimates of area-averaged precipitation rates were also flawed. Which is the truth, ground gauges or satellite retrievals?

My head was spinning. I had lots of things to learn in only a few weeks. First, I knew nearly nothing about the science/engineering we were tackling. For example, what did I know about satellite orbits? I had a hard time learning about this obscure corner of classical mechanics. When I asked a few experts, it was hard to get them down to the level of ignorance they were confronting with me. Of course, Tom had been working in this field for years, and he had a pretty good idea of all the pitfalls. He was a master of crunching arithmetic and order of magnitude estimates in his head, sometimes not mentioning what he was using in his estimates. Not only that, but he is incredibly smart, and he does not suffer fools lightly. I was not sure where I fit in the spectrum of intelligence, particularly the engineering, hardware, and the like. I suppose he looked over my recent record and decided I must be reasonably bright, but that I was a theoretician, and like poets, they can be pretty impractical. He did not know what to make of a scientist who could barely change a tire or repair a leaky toilet float valve.

We did have the commonality that we were both born Southerners (a red neck and a hick), and although most of my ugly Tennessee twang had washed away over the years, his North Georgia accent was probably being enhanced just to drive his surroundings crazy. He drove a rusty pickup truck with a topper roof over the truck bed. Another good thing was that we shared a good sense of humor that brought us through some hard times along the rutty road to getting a satellite mission accepted, launched, and into orbit. To this day. I have no idea whether Tom thinks I am an idiot or just harmless. I suspect he thinks most of his peers are idiots. I hardly had time to brood over it.

So in the late spring of 1984, I set out to learn what I could of these issues. The first thing was to learn how the instruments worked. This was not trivial, because there were so many features (such as what wavelengths in the microwave to use for different applications) and their coordination (again such as what narrow bands of wavelengths to use and what are the peculiar tradeoffs and how to optimally combine data from different sensors). I also had to learn about satellite orbits and how the data is passed from the satellite down to our Earth receiving stations (it is sent to another satellite higher up that relays it). I did not need to know every detail, but in speaking to meteorologists or managers, I had to have some command of these matters, since doubters and competitors love to ask questions, and they delight in stumping you. Through this study, I came to know more about the combined subjects of climate science and the TRMM spacecraft than my audience. We never oversold the mission's capability. When asked about studying hurricanes, I always answered that it might be possible, but we were not promising much there.

Of course, Tom knew all of this, and I was shy about asking him for details because he was so busy with his branch stuff. When I did go in to ask him, I felt the same way I felt when I slunk diffidently in to see Charlie Goebel in graduate school. I would go in, ask the question, hear the incomprehensible reply, and back myself out the door, perhaps knocking a book off the table, while the answerer turned back to what he found more interesting on his desk or out the window. A week later I would have figured it out, pranced in to show him my solution, and he would say "Yeah, that's what I told you." Tom was not as tough or bright as Charlie, but it took me some time to get to know him. I think he at least thought I might be a serious scientist (Charlie said that about me once in 2006). I think it was a consolation for the fact that he thought I was an idiot. Sincerity is good for a scientist, but for a theoretical physicist, it is trumped by pure intellectual power. One wise thing Tom taught me was that as we proceed, we must never present ourselves as being originators or anything of the kind—no bragging, no egos. It turns the managers (who do not share all the charming quirks of scientists) off. Getting the mission to succeed was all we needed to feel good about ourselves. We were in complete agreement. We followed this rule throughout the process.

## Tropical Rainfall Measuring Mission

We named the mission the Tropical Rainfall Measuring Mission (TRMM). This handy acronym came to me from a referendum that was held in Prince George's County, Maryland, on a proposed tax increase. Its name was TRIM, connoting the proposed cut back of expenses in the budget. The name TRMM for the satellite project stuck. I did receive a phone call from Tim Barnett, a friend at Scripps Institution of Oceanography (SIO), with a kindred sense of humor. He said, "Did you know that trim is a slang word on the streets in some cities?" I responded, "Too late to change it now, old friend." I will not reveal the street definition here.

Laura and I bought a Labrador Retriever puppy, black and biggest in the litter, who grew to be a ninety-pounder. His old man was a police dog. His name was TRMM. We hoped that TRMM the satellite would launch while TRMM the dog was still around to help us celebrate the event. The poor fellow was pretty neurotic, among other phobias, he was afraid of water—not a good sign in a Labrador. Once on the drive to Texas in August 1986, he panicked and ran away into a forest outside of Knoxville when we had stopped to visit relatives. We searched and searched, and then essentially gave up, but my uncle called that evening, where we were staying with friends, saying that he had returned to their house. We retrieved the retriever. Sadly, TRMM the dog died of cancer at age seven. It would be another seven years before TRMM the satellite was launched.

My primary role in the advancement of TRMM was in selling the mission through speaking engagements, including at the competitions among contending programs. In addition to the sales work, my group provided research primarily on sampling and other statistical issues needed to prove the feasibility of the mission. I also led in getting things together, such as workshops, symposia, and so forth.

## Competition

In the fall of 1984, there was a small symposium at GSFC to winnow the list down to four proposals that would go to the next level at NASA HQ. I presented our case for TRMM, and I felt that it went well, especially given that there was some pretty stiff competition. We were happy to be selected for one of those slots. I went to HQ about a month later and presented the proposal there against about sixteen competitors. Francis Bretherton hosted (refereed) the contest. By that time, he had moved to the University of Wisconsin, but I knew him well from his days as director of NCAR during my year there as a senior visitor in 1974–75. He always intimidated me with his Cambridge education and English accent—a brilliant and well-informed scientist/administrator (these are rare). We won the competition against some very good contenders. It did not hurt a bit that John Theon, by then at NASA HQ, had been a champion all along for a rainfall mission. Many of our competitors' missions found their way into orbit over the years.

After that big event, we set up to have a presentation with the GSFC director at that time, Noel Hinners, a PhD geologist by training. Much earlier, he had worked with the Apollo astronauts. I made the presentation to him and a few house graybeards at GSFC, and he signed on. That means that the mission had advanced to "Phase A." Phase A here means that money would be available to conduct studies on the feasibility of the mission, including recruiting potential team members from the scientific community and distributing limited grant money (via HQ). "Phase B" meant that we would begin "cutting metal," as the jargon goes. I was designated as the "study scientist."

## Study Scientist

Along with the study scientist, there was to be a "study manager," typically an engineer accustomed to getting all the hardware and all the logistics underway. Later that person would supervise all the contractors that participate and master the various parallel streams of work that would have to converge as we approached the launch date. This person has to be super-organized and tough on contractors and everyone else in the program.

The designated study manager happened to be a new employee with a PhD and just finishing a postdoc in meteorology, but this individual did not seem like such a graduate. Our first meetings were very tense, and the new study manager was very authoritarian toward the scientists involved—perhaps this followed a rough schooling by inhouse experienced project managers. Tom and I discussed the matter, and after a few more such meetings, we decided we had to get a new study manager. It was my job to get the study manager off the project. This person was on six-month probation (as are all government employees initially). I had to go over to the directorate Chief of Space Operations and convince him to get rid of our study manager. Beforehand, I found that our study manager had been educated and trained by friends of mine, and they recommended that I proceed with my onerous mission—the person was unsuitable. I seemed to have convinced the study manager's boss, since I heard a few days later that we had a new manager. This was one of my less pleasant experiences at GSFC. I could not help but wonder what was in store as we proceeded.

Tom and I were starting to react to criticisms about the cost of the mission. We knew it would inflate as we went along. At first it was to be very cheap with the already available ESMR, a visible/infrared radiometer to measure both sunlight coming down and upwelling infrared radiation going out to space. A cheap lightning detector might also be feasible, and its supporters were starting to apply pressure. But we really needed a nadir-focused radar that could do cross-track scanning. Such an instrument had not flown on a spacecraft before. If we included the radar, it would drive up the cost tremendously. Also, to be useful, it would have to have fine spatial resolution (boxes of a few kilometers on an edge to capture individual convective storms), which would mean a low-altitude orbit (only about 350 km above the surface). The next big-ticket item was the rocket launch that would probably cost about $70 million, but it might be possible to share the launch with other payloads being launched at the same time.

We began looking for an international partner. We considered the Germans: Fokker, the aircraft-manufacturing corporation, was at that time involved in some space projects. When that did not work out, we turned to the Italians, who were interested in our using their launch site in East Africa (one of their former colonies) right on the equator. This was interesting, but it eventually fell through. We looked at the Netherlands. They were interested in having a satellite that could be used to detect the theft of very precious lumber (such as ebony) that was being stolen and loaded on illegal ships from other countries in the East Indies. That was going nowhere, but amusing to think about. Finally, we came to the Japanese.

There are many stories about how that partnership worked out initially. Ever since he arrived at GSFC and perhaps long before, Dave Atlas had been interested in measuring rainfall from orbiting spacecraft. It was surely one of his reasons for coming to GSFC and an opportunity to do big things. He and Otto Thiele held a huge workshop at GSFC in 1981. I was so absorbed in other matters that I did not even know about it— or more likely it went right over my head. Participants from around the world were in attendance.

## Japanese Partners

Otto attended and made the arrangements for the workshop, and that is where he recognized the broad interest in the problem, long before he lured me into it. Several Japanese scientists attended the meeting, and Dave invited Dr. Nobuyoshi Fugono to take an official extended visit to GSFC. He did make the visit and afterward returned to his job as a branch head at the Communications Research Laboratory located just outside Tokyo. I heard all about this in 1984. Fugono returned for another visit to GSFC just after my appointment as study scientist for TRMM. He came by to speak with me in my office. After a few minutes, we talked about international partnerships. I popped the question that perhaps Japan might join us. I did not know that he had just come from Dave's office. He did not let on and acted surprised, actually shocked. So I think many persons had popped the question to Dr. Fugono that day. He said to all of us, I am sure, that he would have to think about it after returning to Japan.

By this time, we had assembled a TRMM Science Team with Eugene "Gene" Rassmusson as chairman. By 1984 Gene had retired after a distinguished career as research meteorologist at NOAA. He retired to a research scientist position at Maryland. He was the perfect choice for this job because he had done seminal work on the El Niño phenomenon, and his leadership and personal skills were outstanding. Others on the team included Dr. Edward Zipser, from NCAR, who had been a key player in organizing and leading the aircraft experiments in the GATE experiment—GATE stands for GARP Atlantic Tropical Experiment and GARP stands for Global Atmospheric Research Program, which I will say more about over the next few pages (later, Wilheit and I recruited Ed to move to Texas A&M as department head); Dr. Joanne Simpson, branch head for mesoscale meteorology in GLAS, and perhaps the leading scientist in tropical convection; Prof. Peter Webster, at that time at Penn State, and a super-expert on tropical meteorology, especially monsoons; and of course, Tom Wilheit, Bob Mineghini, and me.

After a favorable but tentative reply from Japan, the TRMM Science Team packed up and flew to Tokyo, except Dave. We were shown the finest of Japanese hospitality in dining and in every other form of courtesy for a host country. Tom, Bob Meneghini, and several others including myself gave seminar-type talks. I spoke about the TRMM plans. Afterward during the break, Prof. Taroh Matsuno came over to ask me if I were the same person that wrote the papers on energy balance climate models. I was pleasantly surprised because he is one of the gods of tropical meteorology and atmospheric wave phenomena. He asked, "How could such a person be so deeply interested in two such disparate matters?" I answered that I am a physicist always looking for good problems to solve. I saw him on many other occasions later, and we always remembered his asking me about my parallel life's work.

It is fun these days to recall how presentations were made in those pre-PowerPoint, pre-laptop, pre-email times. We had to prepare our slides as overhead transparencies. I would make sketches of bullets or whatever graphs or drawings I wanted to show. Then I took them across the GSFC campus to an art shop. They prepared the transparencies with heavy cardboard rims around the margins. If you had twenty slides to show it was quite a pile of stuff to cram into a suitcase. You might get a call to pick them up a week or two later. They had to follow NASA protocols on the art, style, official logos, and so on. Multiple colors in the same diagram were done by cutting out and overlaying transparencies bound all together at the heavy margins.

The other thing I remembered from that first trip to Japan was the short meeting we had with the Japanese negotiators. Their opening gambit was that the United States would foot all the costs of the project. This was clearly impossible, and someone with sufficient authority (probably from engineering at GSFC) said that would not work. Having no experience, I left somewhat bewildered, but we were reassured that this is the way business is opened in Japan, and indeed this interpretation was correct. Over a few more visits, it appeared that we had worked out a roughly 50/50 arrangement. Those higher in the food chain than I had worked out the compromises. I had no part in these negotiations.

The arrangement was that the Japanese would provide the launch with their new H-IIA rocket, and they would build the electrically scanning radar from scratch. The radar concept they proposed had never been flown on a satellite. JPL also wanted to supply a radar of similar design, but that was not feasible because the Japanese would foot the bill for theirs. The H-IIA rocket was a new means of putting satellites into orbit. But from a Japanese perspective, its main purpose in the long term was to provide commercial customers with the capability of launching geostationary satellites (the ones that are about 40,000 km [25,000 miles] up and stay over the same point above the equator; we use these for satellite TV, for example). The first launch of an H-IIA rocket failed, and it fell into the ocean. So we had to cross our fingers. If it failed, we would be out of business. Happily, on November 27, 1999, the launch did succeed, a mere thirteen years from our first presentation.

The American role was to contribute the satellite platform (called the bus), including the solar panels and all the ancillary electronics; the microwave radiometer, called the TRMM Microwave Radiometer (TMI); a visible/infrared radiometer for measuring the Earth radiation budget in the tropics; a lightning detector; and the services of the US satellites that pick up the data from TRMM and transmit it down to stations on the ground. The TMI and the bus were to be built from scratch at GSFC. The final assembly was also to be at GSFC. The assembly phase was critical because some of the instruments, the high-performance connecters, and other components are bought or contracted out. Everything has to be put together carefully in a controlled environment. Finally, the assembly has to be tested before it is put aboard the launch vehicle. Another issue was that it was to be done at GSFC rather than by a contractor. This is done from time to time to keep the government engineers on their toes. It all worked out.

I visited Japan four or five times. It was always a pleasure, but I am not a lover of Japanese food, no matter how nice the restaurant. Our hosts always served one midday Chinese meal, and it was always delicious (usually my favorite of the visit). On one of my trips, I had back-to-back meetings that I had to attend. I flew to Japan, attended a meeting there and promptly flew back to pick up Laura, then on to London. I spent less time on the ground in Tokyo than in the air to and from. In the United Kingdom, we took a car down to Brighton where I was grounded with a head cold that had settled in my eyes. After a day or two, I was ready for my meeting in Reading, where John Theon from NASA HQ was hosting a conference on measuring rain from space.

After Reading we drove up to Oxford, where it was dreary and rainy. A former student of mine was a postdoc there. I had not met her husband, who was a rather famous professor at Oxford. She told us to go to a certain college on the campus and that we could stay in her husband's "rooms." We parked and dragged ourselves and luggage through the drizzle to the college. The "porter" at the door had the key waiting for us. Just to fill out the picture, Laura was several months pregnant. The "rooms" consisted of a pair of austere twin beds, no carpets, squeaky wood floors, and a bathroom down the hall. I said, "Laura, I cannot put you up here," and we dragged ourselves back to the car and found a bed and breakfast just outside of town. Then we called on the former student and took her out to dinner. Her husband was too involved in a soccer game (what is a World Cup?) on TV, so he did not join us.

After giving TRMM promotion presentations at NCAR, the University of Washington, Scripps Institution of Oceanography in La Jolla, California, and McGill in Montreal, we finally passed from Phase A, the study phase, to Phase B, the "cutting metal" phase. I continued to make pitches for TRMM even after we achieved Phase B status. It was thirteen years from the attainment of Phase A status. At any time, the door could have been slammed by Congress or the executive branch of the government. Hence, I continued over the years to give talks to institutions that might help us in a squeeze. These included trips to Tokyo, Taipei, Shanghai, Beijing, Nanjing, Geneva, Madrid, Gerona, Paris, London, Reading, Prague, Tel Aviv, Bangalore, Caracas, and even more venues in North America.

By this time my old friend Marvin Geller was the lab chief of GLAS. Dave had retired and was off to JPL as a distinguished visiting scientist for a year. There was a search to find his replacement. Marvin and I were both on the search committee, and we made one offer, which was turned down when the candidate (professor) in question found that the pay cut involved in moving to the government was just too much for him to bear. Both Marv and I felt that the remaining candidates were not good enough. Instead, we decided to run against each other for the job.

Marv got the job, and I was much relieved. I had fooled myself into getting into the race. Marv was such an eccentric character, and I thought they simply would not select him. Of course, he was very competent and more than qualified professionally, but he liked to jog around the campus in shorts and sandals—I thought NASA brass would never hire such a Bohemian—I would be his backup. From my point of view, the case against me was that, as the study scientist for TRMM, I would not be in a good position as division head to push for my own mission. Marvin did succeed Dave, and one of his first acts was to choose a project scientist for TRMM. He chose Joanne Simpson, and it was a great choice. He told me that I was a good scientist, but probably would not be happy as a project scientist. He was right. In fact, Marv made a great lab chief, and he successfully pushed for TRMM and took it through some close calls.

Aside from being the project salesman, I was in charge of the serious statistical problems for the mission. We had to show that it was at least plausible that we could estimate the average precipitation rate over a month and a 300-mile square to an accuracy of 10 to 15 percent. My first paper, one that I really enjoyed, was with a mathematician by the name of Alan McConnell—the most idealistic and liberal spirit I have ever known.

Alan had been a tenured professor, teaching at Howard University in Washington, DC, for many years. Once in a class on remedial algebra, he scolded his (mostly middle-class African American) students for trying to maintain their schoolwork while working at an outside job. Most of the money they made (he thought) went to entertainment rather than tuition and related expenses. Alan told the fable of the monkey that reached into a jar filled with rock candy. When the monkey grasped too many pieces of the rock candy, he was unable to pull his hand out of the jar.

The moral was that they should not try to do too many things at once. Some of the students took this to be a racist statement, but it was not. Perhaps it was a lack of sensitivity. Alan even had sabbatical visits where he taught in the deepest parts of Africa. Being a rigid man of rigid principle, he refused to teach the class while one particularly obstinate student remained in it. He was fired from his tenured position, and he decided to fight for his job in the DC court. He must have lost more than $50,000 in trying to get his job back. The trial finally occurred years later, and I flew up to DC from Texas and was prepared to testify as a character witness, but I was not called to the stand. Ultimately, he lost his case.

Alan then opened his own one-man contracting firm. I suggested the name "Pixel Analysis," and it stuck. I wrote one paper with him, and it was very important for selling the TRMM concept. We ran computer-simulated flights over a grid box at the ground and used radar data from the GATE experiment in the tropical Atlantic and showed that the sampling was adequate (approximately 10 percent), at least for that location, the only one available.

I had other collaborations that were important in the acceptance of TRMM, especially one with a very delightful Israeli statistics professor at the University of Maryland named Ben Kedem. We referred to the paper as KCN (the formula for deadly potassium cyanide). The C was for our coauthor Long Chiu, whom I mentioned earlier. Chiu and Kedem went on to publish other significant papers related to TRMM. There were also papers coauthored by Thomas L. Bell, one of the members of my twig. We cooked up a model of rain that was unique. Over the years, Bell was the preeminent expert on sampling errors for TRMM and its successors.

Rain is intermittent—most of the time, it is not raining in either time or space (along the surface). A good analogy is this: the rain rate as a function of position is like a bunch of islands poking their peaks above a level of water, which is the no-rain level. When raining, it is distributed not normally, but log-normally. In other words, the logarithm of the rain rate is distributed normally. We had some data from the GATE experiment to tell us about the space-time correlations when it was raining. We found a way to simulate realizations from such a random field. We came up with this idea in discussions. Bell worked out the mathematics and programmed it. He created this peculiar random field that could be used in computer-simulated flights over it to estimate the sampling errors. Bell went on to write more interesting papers using this technique.

Another step I made as study scientist was bringing in Prof. Robert Houze, from the University of Washington to the TRMM program. I already mentioned the review paper I read that he wrote with Alan Betts. The review was based on the analysis of data from the GATE experiment.

The data were primarily radar data but with all the meteorological parameters as well. As I stated earlier, GATE stands for GARP Atlantic Tropical Experiment, and GARP stands for Global Atmospheric Research Program. Let me give a little explanation of these important programs. I mentioned in a previous chapter the rise of interest in large observing programs proposed and carried out by the giants of meteorology in the 1960s, including Jule Charney and many others. The overarching program was GARP. But there was already interest in the tropics as a key ingredient in the global atmospheric circulation. Hence, GATE, a specific field campaign, was planned and carried out in the summer of 1974. In 1975, B. J. Mason, mentioned in chapter five, published a paper in *Nature* that summarized the GATE results.

> The abstract of Mason's paper reads:
> The Atlantic Tropical Experiment (GATE), the first major
> observational experiment of the Global Atmospheric Research
> Programme (GARP) and designed to explore the tropical
> atmosphere, was successfully completed in September 1974. In
> what was probably the largest and most complex international
> scientific experiment ever undertaken, ten nations—Brazil, Canada, France, Federal Republic of Germany, German Democratic
> Republic, Mexico, Netherlands, USA, UK, and USSR—working
> in close collaboration, contributed 39 specially equipped ships,
> 13 large research aircraft, several meteorological satellites,
> and some 5,000 personnel to an intensive three-month study
> of weather systems in the tropical eastern Atlantic Ocean; in
> addition, some 50 countries in Africa and South America participated by making additional surface and upper-air observations.

I had never met Bob Houze until his visit to GSFC in late 1984. GSFC was a routine stop for Houze. He was told (probably by Otto) that he should drop in my office. After consulting with several scientists who knew Bob (he probably was doing the same about me), I greeted him and shocked him by popping the question regarding his possible role in the ground truth portion of the mission. In fact, I wanted him to be the leader of the ground truth program. His immediate response was that he already had an NSF grant and that was keeping him pretty busy. I explained to him the possibilities of my proposal to him, especially the importance of the ground truth program and "while no one is looking" the great science that could be done in running the program. Over the years, if successful, there could be funding for field programs galore. He would not have to leave his prestigious position at the University of Washington. He said he would think about it. I called him a few days later, and he accepted the role. Over the years, he worked closely with Otto in carrying out the program with energy and creativity. Bob's PhD students were many and successful. They included Courtney Schumacher, who is now a Meisinger Award winner and a full professor at Texas A&M. In fact, I think the assignment with all of its potential was key to boosting Bob from a fine journeyman scientist to one of the very most recognized and decorated in his field.

Joanne Simpson was the first female to receive a PhD in meteorology in the United States. She was a great choice to lead TRMM from that point as project scientist. Dave had worked with her for many years, and she had been a leading tropical meteorologist for decades. Now in her sixties, this would be the crowning achievement for her long and distinguished career. Joanne left an endowed chair at the University of Virginia to come to GSFC.

## TRMM and Climate Science

TRMM was a great success based on any metric. The satellite flew for seventeen years before the instruments were turned off. The spacecraft entered the atmosphere and burned up on June 15, 2015. See:
https://phys.org/news/2015-04-trmm-rainfall-mission-years.html

In terms of usage of the data from the mission, we can estimate its impact by the number of times it appears in the title or abstracts of papers. From 1990 to late 2020, "Tropical Rainfall Measuring Mission" or "TRMM" occurred in the title or abstracts of 5016 papers. These papers have been cited in papers over 123,650 times.

TRMM has played an important role in climate science. Many meteorologists recognized even before TRMM that the tropics are the most important of the poorly understood macroprocesses in the climate change problem. The tropopause is the boundary between the troposphere and the stratosphere. It typically is at about 10,000 m (30,000 feet), where the jet airplanes fly. Air in the troposphere is mildly unstable due to convective overturning. The time scale for the vertical mixing is a few weeks. In the tropics, water vapor is plentiful in the atmosphere. When the water vapor is lifted upward for one reason or another, it cools, and as condensation occurs, an additional vertical kick is provided from the heat released in the condensation process. The energy release is enormous—so much so that it basically drives the circulation of the entire system.

This sequence tells us that this convection is an engine that occurs in the form of towering tropical clouds no larger than a kilometer or two in horizontal scale. Clumps of these are found all along the Intertropical Convergence Zone (ITCZ) that encircles the planet near the equator. Over land, this is where the great tropical rain forests grow. The rising air has to fall somewhere, and that happens in the subtropics. This rise along the equator and fall in the subtropics, up to about 30° north and south, are called the Hadley Cells. Intense convection takes place over the tropical oceans. The details and dynamics of these processes are poorly understood in meteorology. In some ways, the storm tracks in the middle latitudes are much easier to understand and predict. Gathering data over the tropical oceans (similar to the polar regions) is logistically difficult, expensive, and dangerous for observers. In order to improve our climate models, we need to pay more attention to the processes in the tropics. We need continuous data streams to tell us where around the ITCZ the convection is breaking out. The statistics of these energy drivers is key to a better understanding of the system's dynamics. At present, our models still do not represent the tropical processes accurately. Part of the problem is the fineness of the resolution of the climate-model grid in the tropics—the convective processes (storms) are at a smaller scale than the model grid boxes.

Empirical data on climate processes improve the models we need. The global coverage of satellites provides this kind of information. This last section provides a very brief introduction to how a satellite system like TRMM helps in solving climate science problems. Finally, it is the iteration between empirical data and numerical weather and climate models that will fill in the gaps. Our climate models improve through this back-and-forth activity. The TRMM satellite was so successful that a follow-on satellite called the Global Precipitation Mission (GPM) was launched in time to overlap for a few years with TRMM.

(Left to right) Jagadesh Shukla, David Atlas, Joanne Simpson, and Jerry.

Line drawing indicating the different instruments of TRMM and their targets along the ground track.

(Left to right) Otto Thiele, Joanne Simpson, Bill Lau, Jerry, Phil Arkin, Bob Adler, and study manager in Japan.

Naboyoshi Fugono (left) and Jerry (right) in Japan.

# CHAPTER 10

# RETURN TO ACADEME

## *Last Stop*

It was time to leave the government and return to my more preferred station in the university world. While our time in the DC area was exciting, the yearning for a smaller town won out. I began to build a group funded by NASA and NSF. Some of my first (accidental) students and postdocs turned out to be among my best.

## Leaving Camelot

Sometime earlier I had taken a short course on discovering and developing leadership at NASA. My score was low. The councilor told me that my test results showed that I would not like management because I cared too much for the employees (scientists) working under me—I would find it hard to be tough on them, and I would take my problems home with me. After thinking about it, I can see that the test results were valid. I am much more at ease in a job where influence comes via ideas rather than power. Leading small groups is far more satisfying than wielding great power—I had reached my limit in the management sphere. I feared that staying in the government would steer me into more and more management. Meanwhile, I was getting attention from universities to return to academia, for which I was more suited. I had a concrete offer from a Big Ten state university, probably with more prestige than Texas A&M, but I was not comfortable with it.

A few scientists in GLAS knew that I was starting to shop around—word travels fast in those cultures. I had thought of leaving in 1984, before I got involved with TRMM, although I am grateful that I stayed for the TRMM experience. One oceanographer at GLAS said that the Department of Oceanography at Texas A&M had advertised for a senior person interested in climate modeling and oceanography. I took a look at the situation, and Otto reminded me of how great Texas was. I was asked to visit and present a seminar. There were two finalists for the position. The other candidate came before me. He was escorted over to the provost's office by the then department head, Jim Scoggins. With Jim sitting beside him he demanded to the provost that he be given the funds for a private supercomputer, plenty of space, and of course, he would be the department head. When I visited, everyone was nice to me. Understandably, Jim did not care much for that previous candidate. My Southern manners during my visit to the provost's office were more suited for the situation.

I presented a seminar about TRMM and my role in it. There was a party at Dean Mel Friedman's house. There were some of the most renowned professors at Texas A&M there—John Fackler, a distinguished professor and dean of the College of Science; Emanuel Parzen, a distinguished professor of statistics; and many others. I got a little tour around the town and noted that real estate was very cheap. They brought Laura and me down to survey the real estate situation with a realtor, and after she had her look and approval, I accepted the job. College Station and Bryan (geographically merged together) were pretty bleak at the time. But over the years, it has evolved into a lovely place to live.

After moving to Texas A&M, I continued to work on statistics problems related to TRMM. This provided support for graduate students and postdocs funded through the TRMM program. I worked with some remarkable young scientists on these problems. My first hire was a postdoc, Kyung-Sup Shin, who had just received his PhD under guidance from Jim Scoggins. I found Kyung-Sup to be a star with great leadership qualities. He was an excellent programmer, and he wanted to work on the TRMM program. I suggested a sampling problem to him, and he proved to be perfect for my purposes, resulting in a major paper we published in 1988. Graphics from that paper were used in every presentation I made afterward about TRMM. Kyung-Sup was a self-starter. He not only worked on problems I suggested, but he took the initiative to engage with other group members on related problems.

Over the years Kyung-Sup returned to Korea and rose to be the director general of the Korean Meteorological Agency (KMA), which is similar to our NOAA, including ships and even military components. When my sixtieth birthday (a big deal in Korea) came along, he brought Laura and me to Korea where I gave some lectures, and we had a wonderful time, especially since at that time there were many TAMU graduates in Korea, and they all attended the festivities. There is a sad end to the story of Kyung-Sup Shin. He was an avid mountain climber, and he perished (probably altitude sickness, but unresolved) on a climb of Mt. McKinley (Denali) in Alaska.

I was blessed with some good students early on, especially my first graduate student at Texas A&M, Ruby Leung. Ruby appeared at my office door soon after I arrived. She said that she was in Texas because her fiancé, Edmund, was a grad student in nuclear engineering. She had majored in physics in Hong Kong, and she had spent some time as a high school teacher there. I offered her a research assistantship and put her in our beginning meteorology grad courses, and she also took the mathematical physics course in the physics department. She was exceptional in all of her courses, and soon we wrote a bunch of papers using a GCM from NCAR—she learned how to use it while on a visit I arranged for her there with another refugee from GLAS, Chin-Ho Moeng.

Before finishing her PhD dissertation, Ruby requested to continue her dissertation research at the Pacific Northwest National Laboratory (PNNL) in Richland, Washington, a DOE national laboratory—Edmund had landed a job there. She finished her dissertation from Texas A&M while at PNNL and has advanced rapidly. Now she is head of a large climate modeling effort stretching across multiple DOE labs, and in 2017 she was elected to the membership of the National Academy of Engineering (NAE).

In my second year at A&M, I taught a graduate course on turbulence. A young Chinese attendee approached me after class. His name was Sam Shin-Poo Shen, a PhD in mathematics from the University of Wisconsin and a visiting assistant professor in the Math Department at Texas A&M. He asked permission to sit in on my class. On the first day of classes I welcomed him, and later we talked. I found that he had taken a directed-readings course from my old advisor Charlie Goebel on general relativity, an astonishing coincidence. Sam and I went on to publish many papers together. After two years and no offer from Texas A&M, Sam took a faculty position in Canada. Later he moved up to be McCalla Professor of Mathematical and Statistical Sciences at the University of Alberta, Canada, and president of the Canadian Applied and Industrial Mathematics Society. Finally he became leader of a large group with the title Distinguished Professor of Mathematics at San Diego State University and also served two terms as department head. He also has a joint appointment at Scripps Institution of Oceanography nearby.

Sam and I worked on a number of projects over the years. Most memorable is a program that started with a collaboration with my postdoc Sho Nakamoto, an oceanography PhD from Texas A&M. Sho and I had done a study of errors in satellite retrievals based on a Fourier Transform approach. We were able to write an equation for the mean square error for a satellite coming over periodically and taking a snapshot of a rain field whose statistics were stationary in space and time. Space-time stationarity means that the statistics, such as mean, variance, and space-time correlation structure are invariant under translations in time and space. The equation for the mean squared error can be reexpressed in the space-time Fourier representation. Because of the symmetries of the sampling and the random field being sampled, the expression under the integral sign can be partitioned into two distinct factors. The first is the so-called design factor depending only on the sampling design (here the repeating visits) and the second factor depends only on the space-time Fourier spectrum of the field. This neat formulation tells us exactly what are the salient ingredients that we have to know about the rain field and what are the salient characteristics of the sampling design—and it does so quantitatively. Later Sam and I took up the more general problem of optimally weighted sampling with arrays of point gauges on the ground and data from point ground gauges and discrete satellite over-flights. We found that the sampling errors due to ground gauges and those from the over-flights were essentially uncorrelated. This meant that the estimates from each design could be combined and that they are independent of one another—a striking result. This is a practice that is often done without even knowing that it should work, rigorously. The estimates from the two separate designs are *statistically orthogonal*.

We conducted other studies that related to the problem of estimating the annual-average and global-average temperature with arrays of point gauges. It is remarkable that one can achieve remarkably good estimates of the annual-global-average temperatures with as few as sixty-four-point gauges. This is because the annual-average temperatures have a correlation length of about 2000 km. If you divide the area of a disk with a radius of 2000 km into the area of the Earth, it comes out to be roughly 64 (~$8^2$). Note: standard errors are usually expressed as $N^{-1/2}$where N is the number of independent samples). Placing gauges more densely than a "correlation length" improves the estimate, but the correlation between more closely packed gauges means the readings from individual gauges are correlated, and therefore the number of independent estimates is less than the total number of gauges. Sam and I are still in contact today and just signed a contract for joint book authorship with Cambridge University Press.

## Global Positioning Satellite Proposals

I was involved in a couple of additional satellite proposals after I came to Texas A&M. These were proposals where I was the principal investigator on projects that made use of the existing Global Positioning Satellites (GPS). This presented an opportunity to use the GPS satellites and show how the signals sent from them can be used for meteorological applications. The GPS satellites are nominally a twenty-four-satellite configuration that is in medium Earth orbit (MEO) at an altitude of 11,550 miles, and they move in circular orbits, whose orbital planes are inclined to the Earth's equatorial plane. GPS satellites send out signals in the microwave range for use in positioning, but these can be used for other purposes as well. We joined with an existing program (COSMIC), headquartered at the University Corporation for Atmospheric Research (UCAR) in Boulder, early in its history (Note: Our proposal exploited the technique of looking from a tiny satellite in low Earth orbit [LEO] to one of the GPS satellites). We proposed to use a constellation of a dozen or so tiny satellites (each the size of a laptop computer) that would receive the GPS microwave signals as the line of sight between the two passes tangentially through the Earth's atmosphere. As one of the tiny satellites goes behind the Earth over a few minutes, one traces a profile of the atmosphere above a point on the surface. The microwave beam is slightly bent as it passes through the air, and this gives clues about the pressure, temperature, and water vapor. This is called an "occultation" measurement, and it has a heritage in planetary science. The technique would deliver about three thousand profiles per day at essentially random locations over the planet—these could be used for many applications such as initializing a weather forecast.

Richard Goody, Harvard professor emeritus and longtime JPL consultant, asked me to attend a small workshop on the radio-occultation problem for Earth at JPL. I did not know anything about the occultation technique, but I knew a fair amount about Richard because of his reputation and having met and chatted with him at a reception some time earlier.

Richard insisted that I serve as principal investigator on the GPS radio-occultation proposal that would originate at JPL. I had a steep learning curve in front of me. Our first proposal was written and submitted. This first proposal was called GPS-Clim and the three institutions (TAMU, JPL, UCAR) were called the GPS-Clim Consortium. Note that the consortium and project title emphasized climate research to make sure it qualified as a NASA project. A weather satellite was a no-no for NASA—weather is NOAA's turf. This really put us at a disadvantage because the main application for our product was in weather forecasting.

So we cast the project as a climate mission. One feature supported by Goody was that this type of measurement would be absolute and not subject to calibration problems. This meant that if we made some measurements now, they could be compared with a similar set of measurements twenty or more years later. Goody knew of my papers on detection of faint climate signals, and these methods were applicable here. Stephen Leroy, one of the coinvestigators at JPL at the time, and I wrote a paper using these ideas and using the results to estimate how the sampling errors could be optimally reduced. Stephen has been a research scientist at Harvard for many years now. Another key player was Robert Kursinski, then at JPL, later at Arizona, who I much respected and who had written several key papers on the physics of the retrieval process. One thing that really struck me was the brilliance, team spirit, and work ethic of the engineers and scientists at JPL. Time was short, and our proposal was rushed. We were not selected, and I thought that might be the end of my role in the program. No, I was hauled in again a couple of years later (1998).

This time we added a new partnership with a Taiwanese group interested in the ionosphere, and a group at UCAR that was led by the president of UCAR, Rick Anthes. I got to know Rick when I was a member of the UCAR board of trustees. The UCAR group had a name for their project, COSMIC. I was again given the title principal investigator (PI), and the project was to be managed by Texas A&M. JPL favored having a university as the local manager with a figurehead PI (me). Having an outsider as PI had worked for them many times. We still had the annoying problem that ours was primarily a weather-related mission. This time we called it "The Atmospheric Moisture and Oceanic Reflection Experiment on COSMIC: AMORE." We had added a few frills that were not yet ready on the first proposal. So this time we had TAMU, UCAR, JPL, and Taiwan as partners. We again did not win, but according to anonymous sources, we scored highest in the competition on science, but the management structure was problematic. I am sure that the partnership caused NASA HQ to cringe a bit because there were problems with the US diplomatic recognition of Taiwan at that time. I also suspect that the relative inexperience of Texas A&M (and perhaps, yours truly) in managing such a large project weakened our prospects.

Radio-occultation work continues through COSMIC, and they have been able to put the tiny spacecraft on many nearly free rides of opportunity. Another piece of good news from all that was that JPL had another proposal that year partnered with a team at the University of Texas, where there is a distinguished center on space physics projects. Their mission nosed us out and has turned out to be a tremendous success. The so-called GRACE mission measures tiny anomalies in the gravitational field continuously over the Earth's surface. An example application is that as ice melts on Greenland and Antarctica, there is a tiny perturbation of the field that can be measured from a pair of small satellites, one following close behind the other. Another success of the mission was that even such faint signals such as changes in reservoirs and groundwater can be measured. After many years of discoveries, GRACE was grounded, but a follow-on mission has recently been successfully launched.

## Big Science in America

Before WWII, a scientist in America was basically an amateur. These investigators often worked in academic institutions, but the hard sciences were constrained by funds needed to support experiments. Any equipment had to be paid for by the academic institution that employed the researcher. Some charitable institutions or individual donors also contributed, but not at the necessary scale and objectivity available through the government. The researcher usually had classes to teach, and research could be only a part-time effort. Tuition fees provided the money for equipment and/or technicians who assisted the primary investigator. In the early days, some states provided funds for research to their public universities. Many scientists (especially physicists and chemists but also meteorologists and others) in America who wanted to learn advanced research techniques traveled to England, Scotland, Germany, Austria, Denmark, Holland, Norway, and Sweden. America was a backwater.

The movement of scientists into American institutions continued even after WWII, primarily because unravaged America had recovered from the war rapidly, and her economy was booming. Returning veterans sought to enroll in universities, expanding the teaching opportunities for professors. Grateful state governments ponied up the funds for the expansion of universities. Such figures in meteorology as Carl-Gustave Rossby, Bernhard Haurwitz, and Jacob Bjerknes came from Europe after the war. Many refugees were displaced professionals in other fields. Later their ambitious offspring grew up in America, attended American universities, and became scientists.

An indication of part of the existence of big science is the participation of the science-related, mission-oriented government agencies. As the Cold War came to an end, DOE labs were flush with scientists and other resources. Many chose to divert funding to climate research. One reason for DOE getting into climate research is their capacity for large-scale computation based upon their expertise in computer simulations for nuclear weapons research. Giant computing facilities are located especially at ORNL and Los Alamos National Laboratory (LANL). The big climate models utilize these machines.

NASA has national laboratories distributed around the country, and each of them has a role in climate science. Some examples include GSFC, Marshall Space Flight Center (MSFC), JPL, and Langley Research Center (LRC). NASA's role is in observing the climate system mainly from satellites but also from research aircraft, balloons, and rockets. There are over a dozen Earth observing satellites currently orbiting, but there are many others that report climate-related data, such as the one called SOURCE, a satellite that makes measurements of the Sun's changing properties; another is GRACE (just described), a satellite that measures the Earth's gravity field, which leads to estimations of the loss of ice mass over

Greenland and Antarctica, as well as many hydrological applications. The upshot is that NASA devotes a sizable fraction of its budget to conceiving, designing, and building satellites. Once in orbit, research is conducted on the retrieved data.

The NSF supports NCAR, the International Ocean Discovery Program (IODP), WHOI, Scripps Institution of Oceanography (SIO), and other facilities that support climate research. NOAA has laboratory and computing facilities spread about the country as well. Other agencies such as the Environmental Protection Agency (EPA) and the US Geological Survey (USGS) have similar missions in climate science. Moreover, there are similar efforts along these same lines in countries all over the world. For example, more than one hundred countries participate in the periodic reportage of the IPCC.

All of these national laboratories and facilities in the United States and abroad contribute significantly to research on climate science. In addition to work done "in house" at the national labs, research grants and contracts are funded for investigators in academia and the private sectors. The annual budgets run into the billions.

This volume of research being conducted around the world indicates that climate science is a member of what I have described as big science. There is an established paradigm that characterizes climate science as a significant scientific enterprise. There are puzzles to solve, and there are thousands of scientists plugging away at solving them. The paradigm of consistency of observation and computation is now well established. Incremental discoveries are made virtually every day. The driving force is the modeling of physical phenomena from prescriptions based on physics and chemistry. The models suggest new experiments, and the observations suggest new numerical predictions. No serious contradictions loom on the horizon. Old believers are few. By old believers, I mean those who still believe that modeling of climate based on physical laws is impossible or worthless. There remain uncertainties in our projections of the future climate, but we now generally understand the nature of these uncertainties. It appears that climate science is in the happy state of *normal science*.

Another indicator of the normal science status of climate science is the number of climate scientists who are winning awards, starting with election to "fellow," from the "old fashioned" societies like the American Geophysical Union (AGU) and the American Meteorological Society

(AMS), both slow to pick up on it, now almost dominated by it. The NAS has published many reports about progress in climate science. Climate science has now passed the early growth phase of a new science. The broader science community recognizes this status.

## Progress in Climate Science

As I was moving from NASA to Texas A&M, general circulation climate modelers had just figured out how to couple atmospheric and oceanic general circulation components of the system together in a unified working whole. It took about a decade to stabilize hooking these two components together without the model going haywire. The problem stemmed from two effects: 1) The atmospheric time scales for change are much shorter than those of the oceans, and at the same time the characteristic spatial scales of atmospheric systems are about 1000 km (about the size of Texas), whereas the corresponding scales for the oceanic motions are an order of magnitude smaller. Getting the numerical models to fit together without the model becoming unstable is not so trivial. 2) The problem was eventually traced to the fact that incorrect fresh water flows from rivers, as well as poor precipitation simulation on the sea and land surfaces, were upsetting the density distribution in the oceans. Once this and other less important issues were fixed by improving the hydrological components of the atmospheric model module of the computer code, the composite model began to settle down and run smoothly.

The sensitivity of the model climate's simulated global average temperature to external perturbations remains a problem. A fundamental property of a particular model is its equilibrium sensitivity to doubling $CO_2$. In this test, the model is run for a long time—it rattles along, but the statistics are stationary, meaning that averages and correlation structures are fixed—then the experimenter suddenly doubles the $CO_2$. After the model settles into its new (warmer) statistical equilibrium state, the new (average) temperature is evaluated. The change in global mean from equilibrium to equilibrium is referred to as the *equilibrium sensitivity*. If we inquire of the world's collection of models, we will find that there is a (nearly normal) distribution of climate sensitivities whose mean is 3.0°C (5.4°F) with a standard deviation of 1.5°C (2.7°F). So why do the models differ so much from one another?

We have a pretty good idea about some of the factors that bring this about. Using a variety of simpler models to learn that if all other things were held constant (no feedbacks), and we doubled the $CO_2$, the climate sensitivity would be 1.0°C with only a 10 percent or so error margin. Something in the system is amplifying the response in our climate models (and in the real world as well).

Theorists have thought for over a century that the main reason is due to water vapor feedback. If an external heating shift is applied to the system, the internal workings of the system will warm the atmosphere slightly, but water vapor (specific humidity: the ratio of the number water vapor molecules in a volume to that of background air molecules) will increase quickly (in a matter of days to weeks). Water vapor is a powerful greenhouse gas. A variety of experience (model and empirical) suggests that the relative humidity (the ratio of specific humidity to the saturation value in percent; saturation is the amount at which the vapor changes to liquid droplets) stays nearly the same, but the actual amount of water vapor in the air column increases with temperature. A simple way of saying it is that the air (more precisely the volume) holds more water vapor the higher the temperature.

So when direct heating is applied to the system, there is an immediate response in the field of water vapor, and this internal adjustment causes additional warming because water vapor is such a powerful greenhouse gas. This is an example of a positive feedback mechanism. Mathematically and physically, a positive feedback is similar to a sound amplifier. When you speak into the microphone that is hooked to an amplifier, your voice is picked up by the electronics, and the circuitry inside the box multiplies the amplitude of your speech (sound waves). If the feedback exceeds a certain threshold, the whole thing can "runaway," and you get that horrible screeching noise often heard in auditoriums where sound reverberates back into the microphone. In the Earth's climate system, this horrible calamity does not happen, but it may have happened on Venus, which could have experienced a runaway greenhouse effect that left the surface hot enough to melt lead. Venus is closer to the Sun and has orders of magnitude more $CO_2$ in its atmosphere.

There are other greenhouse gases besides $CO_2$ and water vapor, and they contribute about half again as much to our greenhouse effect as does $CO_2$. It turns out that we need the $CO_2$ and water vapor and the others, for if we did not have them, the Earth would be too cold for life as we know it. It would probably be a bright shining snowball, and there is geological evidence that this may have occurred a few times on the early Earth.

There are other feedback mechanisms in the climate system. The next larger is probably ice-albedo feedback. To illustrate this case, picture an ice cap at the poles. If the Earth were forced to cool slightly by some external cause (maybe the Sun dims), the planetary temperature would drop a bit. In so doing, the ice cap would grow, and because of its high reflectivity to sunlight, the temperature would sink even further. Hence, the ice-albedo feedback is positive just like the water vapor feedback. There are several other feedbacks, some positive and some negative.

One of the most mysterious feedback mechanisms to modelers is cloud feedback. Clouds reflect back sunlight no matter how high their tops, this property reduces the heating rate of the planet. If cloud area were to grow during a warming, their reflectivity property would act as a cooling agent, and it would constitute a negative feedback. But the clouds also interact with the infrared radiation leaving the planet. They absorb upwelling infrared from below and reradiate it to space from their tops, which are cooler than the Earth's surface, so cloud height also is a factor. In the infrared, the clouds act (somewhat) like a greenhouse. If their area grows, they will act to counteract the effect of the reflected sunlight from the cloud's reflectivity. But if cloud areas grow and their tops become higher, they will radiate energy toward space from a level with colder temperature than before and that will lead to a reduction of heat flow to space via infrared radiation, and the planet will warm.

The solar reflectivity and the absorption of infrared below and radiation from the cold tops of clouds can cancel or go one way or the other, depending on how a model (or how the Earth) characterizes them. Which of these outcomes wins determines the sign (positive or negative) and magnitude of the feedback. It is likely that on some parts of the Earth, cloud feedback might be positive, and at other places it might be negative. The effect is manifested in the details of the general circulation of the atmosphere, where rising air leads to clouds, sinking air leads to clear skies.

General circulation models simulate the flow of the atmospheric air parcels (as a fluid). The way each model completes this part of the scheme is nearly the same from one to another. This conformity is the decades-long heritage of the weather forecasting experience. Models agree on how air parcels move about in the atmosphere at least at large scales. But the precise way water is treated in the general circulation remains a serious unknown because phase transitions in water depend on circulations at smaller scale (for example, tropical convection). Different modeling shops use their own methods of modeling some of these subtle phenomena. They may not matter in a short-term weather forecast but are crucial in the long-term statistics of climate. For this reason, sensitivities estimated from one modeling group to another are not uniform. As of this writing, many investigators from all the participating countries are working to clarify this and other even less understood feedbacks to get to the right answer. I suspect we will not settle this for decades. The major obstruction in solving this fundamental problem is the demand to include more processes, such as vegetation, sea ice, and so forth, in the models before we get the atmospheric model with its water problems under control. This is an internal problem in the modeling community and its leadership.

Bjorn Stephens, director of the Max Planck Institute in Hamburg, recently pointed out in a study that even if we take a planet with an all-ocean surface and no seasons, the models do not agree even qualitatively with the water-related responses to exponentially increasing $CO_2$ over a century. He argues that we must go back to the original plans of Joe Smagarinsky and Suki Manabe to get these fundamentals out of the way before getting too aggressive with a wider range of application of the model results. In my opinion, until we do that, the climate sensitivity problem will not improve (Stephens and Bony 2013).

It would be wonderful if we could solve the problem using empirical arguments and leave the models out. But there is no straightforward way to extract the answer empirically. We can only hope to use data from a variety of sources to improve or at least constrain the uncertainty in climate sensitivity from our models. There is simply never enough data for the empirical approaches, and never enough computer time for the modeling path. But they go back and forth checking each other, looking for inconsistencies in the web of climate science knowledge. The working paradigm assures constant confrontations between data and prediction. The overarching program stumbles slowly but satisfactorily along with what I would call educated trial and error. One thing is clear and universally understood: increasing the concentration of $CO_2$ and other greenhouse gases will increase the planetary temperature. The uncertainty in the exact amount of warming is smaller than the global warming signal. Chapter fourteen will cover the climate projections for the twenty-first century.

## Estimating Climate Parameters

My talents and knowledge are limited in meteorology, but I have more than a common intuition and knowledge of the statistical aspects of climate science. While I have never taken a statistics course, I did have a major in mathematics as well as in physics as an undergrad, and I took a math minor (two years of grad math courses) for my PhD. The standard graduate physics curriculum provides enough statistical challenges to develop a good statistical intuition that has stood me well in climate science. It helps that I love learning about mathematics and statistics, especially from the historical masters who wrote about it so clearly.

While at GSFC, Bob Cahalan and I had written a short but elegant paper showing that we could solve (without a computer) an energy balance model for a planet that is uniform over the sphere in topography, all land or all ocean, but not in the case of some land and some ocean (broken symmetry). Rotational symmetry of the surface geography is essential in solving the problem with well-known methods of mathematical physics. We found that the solution could be expressed as a weighted sum of functions called the empirical orthogonal functions (EOFs). For this type of highly symmetric planet, the EOFs were just the spherical harmonics (well known and useful in classical and quantum physics, for example, the solution for the hydrogen atom that sold the quantum theory), an astonishing result. Before submitting our paper to a journal, I decided to send our manuscript to my Russian friend, Lev Gandin. One never knows whether such a paper will get through the filters in the USSR, but after about a month, I did hear from him. Gandin loved our paper, but he told me that Alexander Obukhov had already solved the most important step in obtaining the result. I had met Obukhov on two different visits at the Institute for Atmospheric Physics in Moscow, where he was director. Obukhov solved the problem in the 1940s in his PhD dissertation under the superstar mathematician Andrei Kolmogorov. This time Obukhov scooped me in 1945 just as Poisson (on the Poisson Summation Method) had two hundred years earlier. At least this time it was only forty years earlier.

By this time, I was at Texas A&M and trying to come up with a good master's thesis project for my new and very promising student, Ruby Leung. I decided to apply something called "information theory" to the same toy model that Bob Cahalan and I had solved. Information theory was invented by Claude E. Shannon in 1948 in a landmark paper titled "A Mathematical Theory of Communication." I was poking around in the mathematical statistics books and came across Shannon's theory, which seemed to fit our problem as well. I found a slightly different method called *transinformation* (Kullback 1959), and it was an even better way to characterize predictability in climate. The simple formulas from our climate model and the simple formulas for *transinformation* fit together perfectly. I never quite figured out Shannon's theory of information and how it applies to digitally coded messages, but I could understand the concept of entropy of information from my physics background. I was surprised that Ruby quickly mastered that old paper that Bob and I wrote as well as the Shannon stuff. We wrote a short paper and submitted it, and it was accepted and published quickly (Leung and North 1990).

I tried shopping the theoretical idea around to interest the weather forecasters I knew at the National Weather Service, but my advice was not followed. I think this was because the connection of entropy and information is abstract and a hard sell to practitioners. Years later Jagadish Shukla and his colleague Tim DelSole rediscovered the possibilities and were surprised when I informed them of our earlier work. Our paper had sat in the literature for many years, and finally, some mathematicians working in climate prediction at about the same time as Shukla and DelSole came up with the same idea. I was delighted to let them know that we had published the idea some twenty years earlier. At last a mathematician named Richard Kleeman published a review paper in 2011 in the journal *Entropy*, and we got credit for being the first to apply this concept to climate predictability. Neither Ruby nor I have done anything further on the subject. It was fun to inform her via email that she was now famous for work she had done twenty years before. Recently I was asked to be a guest editor for a special issue of a journal devoted to climate predictability and information theory. Sadly, I had to decline the offer because my plate was overloaded.

## Playing with General Circulation Models

Partly in order to better inform myself in big models, I cooked up a project of running a low-cost general circulation model (GCM) at Texas A&M. I did not know exactly how to do it, but I knew an old friend at NCAR, Chin-Ho Moeng, wife of my coauthor at GSFC, Fanthune Moeng. I called Chin-Ho about how we might get one of the old NCAR models to run. She agreed to teach Ruby how to run the old CCM0 model. The Community Climate Model number zero (CCM with a number after to denote the evolving generations of the model, this one was the "zeroeth") was created by Warren Washington, Robert E. Dickinson, Akira Kasahara, and their team in the 1970s. Ruby spent a month or so visiting NCAR, and she brought the computer software back to Texas A&M to run on our supercomputer. It was a success, and Ruby along with a few colleagues in the group got it running, and we published a bunch of studies with it.

The planet we experimented with was dubbed *Terra Blanda* because it was to be a very dull place; it ran at perpetual equinox (no seasons), was all land, with no topography, and a moist surface. It had clouds, precipitation, and water vapor that was transported around by the winds. My main purpose was to show that the surface temperature field was very similar to that of our energy balance climate models.

## Detecting Faint Climate Signals

There are four externally induced perturbations of the energy balance: 1) variations of the solar input (such as the faint eleven-year solar cycle); 2) the screening of solar radiation due to aerosol particles in the lower part of the atmosphere (troposphere: roughly, below where the jet planes fly); 3) screening of solar radiation due to volcanic debris left in the stratosphere after an eruption; and 4) a change in the concentration of long-lived greenhouse gases. Over only a few decades the responses to these drivers are all very faint compared to the natural variability even at the global average. Hence, in the 1990s interest rose in using statistical methods to "detect" these weak "signals" in the system. This became known as the climate signal detection program.

If we have plenty of computer time, we can find the space-time signal pattern of the response for each of the four signals using climate model simulations. Then we rig up a way to look at a real data stream to see if the four signals are contained in the natural variability. The results have to be statistically significant (i.e., the result would only occur by chance in a very small percentage of such trials, say 5 or 10 percent). Another way to put it is that the method has to detect the signal patterns well above the noise level. This can be done by optimally weighting parts of the data stream. In order to do the problem, one needs a long time series of data. The surface temperature field, where we have over a century of data, is the best candidate for detecting its space-time signal pattern. Model runs provide good first guesses about the signal patterns, but not their strengths (because of the unknown magnitude of the several feedbacks), which depend on climate sensitivity for individual models. Once we have the signal patterns, we can construct an optimal "filter" through which the data can be run to filter out as much of the noise as possible. These are part of the methodology I learned while working with Sam Shen. The statistical optimization methods are abstract and complicated, but doable. Similar detection methods are used by engineers (and I read some of their books) in all kinds of applications: an example is the signal processing circuitry in the cell phones we carry about at all times.

Dr. Kwang-Yul Kim, a young research scientist working with me at Texas A&M, and I wrote two papers showing how the detection project could be conducted for a single signal and for the largest scales of the temperature field. We used our energy balance model as an example, because one could easily see how the procedure worked in this simpler context. Both our papers showed how the theory of the detection worked, and the second showed numerical results (North et al. 1995; North and Kim 1995). At the same time, Klaus Hasselmann, then director of the Max Planck Institute in Hamburg, was working on the same problem. We submitted our papers at about the same time, and his sailed through the reviewing process, but ours took months longer, so he generally received first credit for his paper. Klaus is an admired friend, and he fully understood that our work was simultaneous to his—his acknowledgment was enough for me. Scooped again, but this time only by months! I was gaining on the mob!

After that I supervised two outstanding PhD students, Mark Stevens and Qigang Wu, who wrote dissertations on the detection problem. Afterward, Stevens took a postdoc at the Canadian Climate Center and later wound up at NCAR. During a time of budget shrinking at NSF, he wisely became interested in microbiology. He took the requisite under-grad courses at the University of Colorado and leaped to a postdoc posi-tion (skipping the PhD in biology) under a microbiologist. He is now a researcher in microbiology nearing retirement at the University of Colo-rado Hospital in Denver. After writing a great dissertation, Qigang Wu had a series of postdocs in the United States and is now a full professor in atmospheric sciences at Fudan University in Shanghai.

The science of estimation theory includes the detection problems in estimating climate sensitivity from a combination of models and data streams, estimation of climate predictability for noisy data streams, and estimation of precipitation from various data sampling designs such as satellite and/or discrete gauge assemblies. Applying these techniques in climate science has been one of my themes for my own work and for men-toring students. I especially enjoyed students and postdocs that were so excited about our work that they read the literature and would drop by my office to ask, "Have you seen this new paper?" The students and postdocs who worked with me on the estimation problems both at GSFC and Texas A&M were highly motivated and a joy to work with.

Former students of Jerry's and of Jim McGuirk's celebrating Jerry's sixtieth birthday in Seoul (1998). Kuung-Sup Shin stands behind Jerry.

(Left to right) Sam Shen, Ilya Polyak, Jerry, and mathematician Rodolfo Bermejo.

204

Jerry at the Great Wall in China.

Backyard party in Bryan, Texas. Front row: Joseph Yip, Cecilia Yip, Yuhong Yi, kids, Laura, Martha Kirchoff. Back row: Kwang-Yul Kim, Phil Riba, Bill Hyde, Sho Naka-moto, David Short, Jerry, Yoo Shin Ahn. Middle row, leaning: Kyung-Sup Shin, Wan-Te Kwon, Ruby Leung, Edmond (Leung's husband).

# CHAPTER 11

# TEXAS A&M UNIVERSITY AND ACADEMICS

## Why Texas A&M?

I enjoyed my time at GSFC. While there, I felt that I was at my peak for creativity and productivity in my job. I had wonderful colleagues around me, and the postdocs and contractors that worked with me were talented, hardworking, and cooperative. No scientist gets a civil service job in that laboratory without four or five years as a graduate student and another two or more years as a postdoc. The salaries for top levels fall below those in typical academic jobs. If the dedicated civil servants I worked with are representative of those throughout the government, then they deserve far more credit from the American people, who are so often misled by cynical politicians and uninformed followers.

While at GSFC and having TRMM off my back, Laura and I began to think about moving. We settled on Texas A&M and here is why. I am a Southerner, and I like Southerners, their manners, and customs, except for some nutty attitudes that I could not stand. Texas A&M was late to wake up as a research university. Until the 1960s, it was all male, all white, and all military cadets. It took a general to change it, a bit like Nixon opening the way to China. When he became president of the university in 1963, General Earl Rudder, who had grown up in Texas and was well received at A&M, had a chance to change the university's culture. He managed to change the all-male student body, integrate the racial composition, and reduce the dominance of the military on campus. He even foresaw A&M becoming a research university. His political connections (Lyndon Johnson was a friend) brought in some federal money for science. It was a start. But beyond his great beginning, he had no experience in the university culture or with aspects such as the notion of academic freedom and the strange diversity of the university community. He died in office at age fifty-nine, still a hero here. A&M remained relatively backward even in 1986, and so was College Station/Bryan. It is only in the last twenty years that it has risen to be—nearly—shoulder-to-shoulder with the University of Texas. We took a big step forward when Ray Bowen was president, followed by Robert Gates, who was great at following up on Bowen's plan, but had to leave us too soon. One thing I liked about both of these men was their notion that a great university has to be great in the humanities as well as in the trades such as engineering, agriculture—and maybe even the atmospheric sciences.

On the third trip to Bryan/College Station, we bought a big old house with about thirty-five post oak trees in Bryan, an older town that abuts College Station. Our house was just down the block from the dean, we were sandwiched between two professor homes, and the house across the street was an academic family. The rest of the neighborhood was similar with mainly professionals. There were a fair number of blacks and Hispanics in Bryan, hardly any in College Station. There were items in the local paper about academic parents overly pressuring the school board in College Station, and I thought integrated Bryan schools might be better for us. We stayed in the same house in Bryan for twenty-eight years before downsizing to a new but modest house in College Station.

## Texas A&M and Meteorology

Soon after arriving at Texas A&M, I began to see how things worked on the campus. The Department of Meteorology was filled with good faculty members, but they had fallen on hard times. Most had come from celebrated graduate programs such as the University of Chicago (P. M. Das), UCLA (Harry Thompson), Colorado State (Jim McGuirk), and Wisconsin University (Dennis Driscoll). Our state climatologist and full professor, John Griffiths, was schooled in England, spent time in East Africa putting together data for a world database, and was well known internationally. Dusan Djuric had a PhD from Belgrade and a postdoc in the prestigious group in Norway. Others were similarly prepared for research careers.

The College of Geosciences was formed when the eminent Horace Byers came as its first dean from the University of Chicago. He was our first (and last) member of the National Academy of Sciences. Horace Byers was a radar man, and he sought to make A&M a great radar center.

He hired Vance Moyers, also a radar man, who became head of meteorology, and I think he brought in Jim Scoggins, who had been a branch head at NASA's MSFC. Jim's specialty was boundary layer meteorology, and he had been very active in balloon work at NASA's Columbia Balloon Facility at Palestine Texas.

When I arrived, the department had settled down into a teaching department, whose main mission was to train US Air Force (USAF) officers to be good weather forecasters. The program brought in officers with undergrad degrees in science for one year to take our upper-level undergrad meteorology courses. They stayed on the campus two summers and the academic year and could add a BS degree to their resume. There must have been about twenty of them coming through each year. The USAF also sent graduate students to work on MS and even PhD degrees. While the faculty made all of these students work hard, very little came out in published results. One reason was that the USAF did not seem to care about scientific publications. Each year about twenty to thirty BS degrees were awarded; their primary education was intended to prepare them for the National Weather Service or equivalent positions in the private sector. In terms of research, there was only one grant worth approximately $100,000 a year in the department. I think there was only one US civilian PhD student in the department, and he was the local weatherman on KBTX-TV. He eventually took a job in public relations and left the A&M program without his degree.

Much of Moyer's time must have been spent in arranging and preparing for a new building to house the departments of meteorology, oceanography, and geography that opened in 1973. It is a wonderful facility of fourteen stories and an iconic radar dish on the roof. The dean's office is on the second floor. The only negative about the 260-foot building, said to be the tallest between Houston and Dallas, is that it has the stove-pipe layout with faculty members from the three departments stacked atop one another, preventing many casual conversations in hallways, only limited brushes in elevators. You might go years without seeing a colleague from a sister department. Nevertheless, the building had an observatory on its fourteenth floor. The observatory is a large room with a 360-degree view; sadly, the room has no elevator service, and later this precluded its use because of its inaccessibility by disabled persons—three wasted floors.

A celebratory symposium was held when the new building was

dedicated in 1973. Speakers for the symposium included the greatest meteorologist of the time, Jule Charney from MIT; Robert M. White, administrator of NOAA; Karl Turekian, Yale University (and a member of my Hockey Stick Committee in 2006); and many others. TAMU's very able former president, Robert Gates, said to me of our building a few years ago that it looked like a fine piece of "Stalinist" architecture (recall that Gates is a Russian scholar and former CIA director—he left us to become Secretary of Defense). OK, even if it reminds me of USSR structures, I have loved these digs—my office for the last twenty-two years has floor-to-ceiling windows on the twelfth floor, facing north with the city of Bryan on the horizon, and there is virtually no glaring sunshine flooding in.

Besides Moyer, there were a couple of additional radar professors in the department. One was a pleasant and spritely septuagenarian who had published a total of fourteen papers in his career. The second was a former PhD student of Moyer's who I heard had received his tenure/promotion without consultation of the faculty. A very nice fellow, he passed away in his sixties with a total number of publications in the single digits.

By the time I arrived, the cozy deal with the USAF students had basically ruined the department. Oceanographer and later Distinguished Professor Worth Nowlin and Dean Mel Friedman urged me to take a joint appointment in oceanography. Mel said it was possible that meteorology might not make it. Indeed, a good friend of mine in oceanography later told me that he asked a distant meteorologist friend of his how many meteorology departments there were in the United States. The friend told Dave that there were probably about twenty. Dave then asked, "How do you think the one at A&M ranks?" The friend answered, "Definitely not in the top twenty."

If I intended to build a great department, I had my work cut out for me. Mel wondered if I would like to be the department head. I said I did not think so yet—perhaps later. I figured I could work with Jim Scoggins, who had a good career and would understand what we could do here. My premise from the start was that since most of the faculty had all been educated at first-rate institutions, it might be possible to work with them. My strategy was to first treat them with respect—always friendly and supportive of their situation. The fact was, they were stuck in the position of teaching two and three courses every semester and even in the summer to get those USAF BS students through the program. The relatively small meteorology and geography departments were the teaching departments, and the other big departments with huge faculties were the research departments. Indeed there were some fine scientists in those departments. So I set out to improve meteorology.

Jim Scoggins excused me from teaching the first year, and I was given an administrative assistant, Martha Kirchoff. Martha had worked in the department for years, and she had a deep understanding of the university and quite a few contacts. I was also given a twelve-month appointment. I started a thing called the Climate System Research Program (CSRP). I built up a large group with funds from NASA, NOAA, and NSF. At one point, I had seven postdocs and five or six graduate students. I had students and postdocs from Japan, Taiwan, Hong Kong, Mainland China, Korea, India, and of course the United States. There were a number of Korean graduate students in the department. Over the years I cochaired two PhD students in statistics and three in civil engineering (water resources). Over the long term at A&M, I supervised one MS student in physics, and I supervised (or cosupervised) dissertation students in each of the four departments in the College of Geosciences. Laura said it looked like the Southeast Asian Embassy. We had Texas Bar-B-Que parties at my house, and every time a student graduated, the whole group had a big evening party at a local pizza dive. Martha guarded my door from casual visitors, sometimes more than necessary. My door was always open, even later when I was department head.

I helped the Department of Civil Engineering (CE) in hiring Juan Valdes, whom I had met at a hydrology meeting in Caracas where I was promoting the TRMM program. I was visiting A&M before I had formally joined the faculty and the department head of CE called me to ask if I would speak with Juan during his visit at the time in College Station. I did, he was hired, and we became fast friends, working together on hydrology applications for TRMM data. Unfortunately, Martha, an elderly native Texan, was wary of all those foreigners roaming around the tenth floor where my group was now dominating. Come to think of it, she was wary of all of our employees and students, foreigners or not. I ran the group as I did at GSFC, they worked for me, but I allowed lots of freedom in their behavior and work hours. Frankly, it worked well for us, but it was stressful for Martha, who was not so accustomed to dealing with eccentric and sometimes weird folks.

It was especially tough on Martha when I hired a senior Russian émigré named Ilya Polyak. My old friend from Leningrad, Lev Gandin, who had by then immigrated to the United States, recommended Ilya to me. I had actually met briefly with Ilya at the MGO in Leningrad in 1977. Ilya was about my age, perhaps a year older. He was a statistical climatologist at MGO in the group headed by my friend Lev Gandin. Ilya applied for a position in my group at Texas A&M University. Lev recommended him strongly, and so did some other friends such as Konstantin "Kostya" Vinnikov and Michael Fox-Rabinovich, both working at Maryland at that time. I offered Ilya a job in my research group at Texas A&M. Ilya had a life filled with hardships, beginning in Leningrad during WWII. Leningrad was under German siege for nine hundred days. Many people starved, but the battered city lasted through the horror. Ilya was one of the lucky children who were shipped out to central Russia partway through the siege. Ilya wrote a memoir in Russian, years before my meeting him, about his early life. It even received some high praise among Russian critics. He eventually came to Moldavia and finally back to Leningrad where he finished his education with a PhD at the University of Leningrad. He was thoroughly grounded in physics, mathematical physics, and climatology.

While at MGO, just as the Soviet Union was changing, his beloved wife died from cancer. This had a terrible effect on him. Depressed, he sought to emigrate to Israel. The rules for such passages were complicated, and I never quite understood them. He remarried and his new wife was able to get to Israel where she lived in a kibbutz. Ilya somehow made it to the United States. He was able to visit her a few times, but the marriage failed. He had a lovely daughter from Leningrad who accompanied him to Texas A&M where she became an undergraduate and eventually graduated.

Ilya was a contentious scientist and would get into heated arguments about scientific matters. Sometimes he was wrong in his argument, but it was hard to get him to concede his point. Occasionally it would become embarrassing when he would get into it with someone visiting my group from outside. But mostly he was sad about his predicament. None of this prevented him from working really hard, and he worked through some very nice projects. Although he was paid rather well as a research scientist, he wanted more money, and I was tapering down my research program at the time because I was anticipating stepping up to be department head. I suggested that he apply for a senior postdoc position at GSFC. He won the fellowship and left for Maryland with his newly graduated daughter. This could be the end of the story, but there was more to come.

I had recommended a book he had already published in the USSR to be translated into English. He wanted me to be a coauthor, but I declined because all the work was his. I contacted my friend Joyce Berry, who had been the editor for the paleoclimatology book that I coauthored with Tom Crowley. They accepted the book, and it was published while Ilya was at GSFC. A review of the book was published somewhat favorably, but the reviewer said the book was rather eccentric—it retained a distinctly Russian style in the writing as well as the figures. I thought the review was OK, but Ilya was gravely insulted and wanted to pick a fight with the journal that published the review. I recommended that he refrain, but he would not, so it ended up with his harboring some bitterness toward me.

A second event took place while he was at GSFC. We had published a series of papers on comparing data with some GCM simulations. The papers showed that there were considerable errors in the simulations compared to climatological data. This was hardly a surprise because at the time the models were still finding their way and improving. Our final paper on the subject was accepted for publication. Ilya received the galley proofs and made some changes. In the last paragraph, he changed the wording to say that climate models were so bad that they could not be trusted for predicting future climate or for use in paleoclimate research. The publication appeared, and even before I could see it, I received an email from Richard Lindzen, who said something to the effect that while he had some sympathy with this language, it seemed much too strong. Of course, I was stunned.

My response was to write a "comment" to the journal correcting the impression left in the last paragraph of the original paper. My one-page comment was published. Many friends expressed sympathy for my reaction. One said that it was the first time he had ever seen a "comment" on a paper written by one of the coauthors. Although he was well-liked and did good work at GSFC, Ilya did not get renewed for his fellowship, probably because of a lack of funds. Ilya left the field of statistical climatology and moved to the pharmaceutical industry where his statistical expertise was in demand. I think he found such employment. I would hear gossip about Ilya through the émigré network, but I never spoke to Ilya again, even though I always sent best wishes back to him through our mutual friends.

The end of the story came in 2016, when I received a phone call from Kostya Vinnikov, saying that he had just received a message from Ilya. Ilya told Kostya that he was dying and that he wished Kostya to tell all of those whom he had offended that he was truly sorry for the way he had acted those years ago. He also told Kostya to not divulge his whereabouts or any communication addresses. And, of course, this decent man, Kostya Vinnikov, did as he was asked. I recently found the obituary online from Massachusetts.

## Beginnings of a "New" Department

I saw from the beginning that upcoming retirements would provide an opportunity to rebuild. I think I had a mandate from the administration to make some changes in the department. Our first hire was Lee Panetta, who had a PhD in mathematics from Wisconsin, followed by postdoc positions at GFDL in Princeton with Isaac Held and later at the University of Washington with Mike Wallace. The next year I was able to bring my partner in the TRMM program, Tom Wilheit, into the department as a full professor. Wilheit was sufficiently well known that his joining us brought very favorable attention. Jim Scoggins was approaching retirement after a decade as department head.

The dean again asked if I were interested, and again, I said no, unless we cannot find a star from outside. We actually had three outstanding candidates who interviewed, and we were able to hire Ed Zipser, a former division head from NCAR. I had encouraged Ed to apply since both Tom and I knew him well from his service on the TRMM Science Team. Ed's claim to fame was his leadership and management of the aircraft program in the massive international GATE experiment that gathered precipitation and other data from ships in the Tropical Atlantic in the early 1970s. He passed my important criterion: Think big! We were overjoyed when he accepted and came to Texas A&M. He spent six-plus years as department head and at the same time developed a strong research program, attracting and mentoring outstanding graduate students of his own. This interval saw the department's reputation grow; the graduate program continued its growth as well. Our next hire was John Nielsen, who was a synoptic and mesoscale meteorologist. I played a small role in hiring him by taking him on a Saturday morning for a lunch of chili and tamales at a very rustic restaurant in old Bryan. John turned out to be a polymath, brilliant, hardworking, honest, and kind. He modified his name to Nielsen-Gammon after marrying a Texas girl, Beth. Wilheit's and my research programs connected with TRMM flourished during Ed's tenure as department head.

Before and during this time, I served on the Board of Atmospheric Science and Climate (BASC), a part of the National Research Council (NRC). John Dutton from Penn State was chairman, and I was informally his deputy. Our great task was to write a report called "Opportunities and Challenges for the Atmospheric Sciences in the 21st Century." John and I along with NRC staffers (especially Bill Sprigg) devoted lots of time to raising funds and holding briefings from many meteorologists and climate scientists. The Clinton Administration was just beginning. Based upon previous reviews in astronomy and chemistry, we thought we could get funding from James Baker, an oceanographer, recently appointed as administrator of NOAA. But even with Al Gore as vice president, NOAA was under budgetary pressure from the White House, and we ended up conducting the study on a shoestring. We persisted and finally got a good report out, but it could have been better had we been able to bring more voices into the process. The report and the many new contacts I made across the entire field of atmospheric sciences helped to bring needed attention to our department at Texas A&M.

I became department head in 1995. At the time, Prof. James McGuirk had been assistant head under Ed, and he continued in that role with me. My time and effort as head included many trips to workshops, scientific conferences, external reviews, science meetings for TRMM, and most interestingly, I was elected to the board of trustees of UCAR, which manages NCAR and other large research efforts that benefit research across the broader academic community.

I learned more about how big science is managed and funded by this experience. Professor and renowned synoptic meteorologist, Richard "Dick" Reed from the University of Washington was on the board with me. At times Dick and I together felt alone in our old-fashioned academic views of what UCAR should be. In our mind, UCAR was becoming more and more a corporate colossus with not enough regard for its then sixty university participants. This perception grew as more and more board members were coming from outside the realm of academic researchers. The management sought more university and corporate administrators instead of professor-types. Of course, well-meaning professors are often naïve about such matters. Our experience as cottage-industry scientists does not always fit today's big-science environment. Notwithstanding the last remarks, UCAR President Richard "Rick" Anthes excelled as intellectual and practical leader of UCAR. In hindsight, I think Rick understood these processes and trends far better than I did.

Soon after this gig, I served for six years on the board of trustees for another organization of the same kind: the University Space Research Association (USRA). Whereas UCAR had as its primary sponsor the NSF, USRA derived most of its funding from NASA. I was a little more hardened to the nature of such nonprofit organizations by this time and the term "well-meaning professor" was an understood term in conversation at board meetings. USRA had employees who worked onsite at NASA centers. It also had other responsibilities such as running postdoc programs for GSFC. Its most interesting role to me was its management of the Lunar Planetary Institute (LPI) in Clearwater, Texas, just south of Houston and near the Johnson Space Flight Center. The LPI is the counterpart to NCAR, but much smaller. Currently, USRA manages the giant radio telescope at Arecibo in Puerto Rico and the infrared observatory, SOFIA, which features a modified 747 aircraft that has doors that open from its roof to expose an infrared telescope that can look into space. SOFIA takes measurements in infrared astronomy while flying in the stratosphere of our planet.

During my time on the USRA board, I served several years as chair of the audit committee. I took it seriously and read lots of literature on how to conduct such an oversight assignment. The audit committee made use of an outside firm that periodically audited the workings of the corporation. There was also an internal officer at USRA HQ whom I worked with in approving the procedures used in internal audits. With all of these boards, a lawyer was involved at meetings of the boards of trustees, and business was carried out formally. We had onsite visits to all of the NASA centers including Marshall Space Flight Center in Huntsville, Alabama; Langley Research Center in Langley, Virginia; Ames Research Center, Moffat Field, California; Glenn Research Center in Cleveland, Ohio; and LPI in Clearwater, Texas. We also had several visits to an aircraft facility at a site near Waco, Texas, where the 747 aircraft was being modified for the infrared telescope. We climbed about the aircraft, peeking at the interior of a 747 that had been stripped such that we could see its ribs bracing the exterior walls.

The responsibilities of a board of trustees include selecting and firing the president. The board does not meddle in the day-to-day operations of the company, but it does have a look at the overall plans, especially the financial matters such as budget forecasting and similar matters. It is the board's job to see to it that these nonprofit organizations serve the public as well as their client universities. While I was on the USRA board we actually fired the president. A large project was running behind schedule, and the president decided to negotiate some terms with the contractors in charge without informing the board. There were other issues that I was unaware of, and I was just rotating off the board at the time. As I understand, the board hired a new president (a retired general, but with a PhD from MIT) with an annual salary exceeding $400,000. To be fair, I learned that the new guy really did succeed as manager, was well regarded, and certainly rescued and expanded the empire.

My very dear friend Alex Dessler (father of my colleague, Andy), then department chair of space physics at Rice University was among the founders of USRA and was appointed its second president (1975–81) while he continued at Rice. He was paid a small honorarium for looking after things. So much for "well-meaning professors."

I also served 1999 to 2002 as a member of the board of trustees of the National Institute for Environmental Change (Department of Energy), the last two years as chair. This service included visits to Argonne National Laboratory and ORNL where relevant research was conducted. I mention these site visits as part of my experience with a variety of government laboratories where climate science was featured.

## Distinguished Professors

When I was interviewing for the job at Texas A&M, I was puzzled that I was sent to talk with several individuals around the campus. About a month or two after arrival, I was told that I would have the title "Distinguished Professor" (DP). I learned later the purpose of those interviews around the campus was to check me out. It is a tradition of the university that an offer of DP should not be part of the job offer. Later this rule was formalized, such that election to the title should depend at least substantially on work done at Texas A&M.

It was very nice to be a DP, but in the beginning, I did not have much understanding of what it meant, except that I would be given a bursary of $5000 per year in "green money." Green money comes from outside donors in the form of endowments, and the interest returns can be spent on pretty much anything, including alcohol while entertaining a visitor. Money from the State of Texas or from tuition and fees cannot be spent on alcoholic beverages. The green-money bursary made me popular with colleagues who had visitors from out of town. In fact, I was often invited out to dinner, where, of course, I picked up the alcohol tab. Another perk was that I got a special parking tag that allowed me to park in any ordinary lot on the campus. I later found that I could even park next to the stadium at sports events. Actually, DPs are often on committees that might meet miles away from their office and the flexible parking permit is a necessity. As time went on, the group of DPs grew and grew—so much so, that the bursary kept shrinking until finally, they gave us old-timers a one-time payout rather than annual bursaries, although newly elected DPs received a bursary for a few years. OK, it was good while it lasted.

Some of the characters in the early days were a little over the top, demanding good seats at football games and other outrageous perks. I thought this was preposterous. I started (privately) referring to the DPs as the "puffy distinguished professors." I served three three-year terms on the Executive Committee of DPs (ECDP), a group of about a half-dozen members elected by the whole body of DPs (at that time numbering about twenty-five, now over fifty). We met monthly with the provost and occasionally with the president. When Ray Bowen was president, he sometimes had the ECDP to the presidential mansion for breakfast.

When George W. Bush was elected governor, he appointed a number of prominent Texas leaders to the board of regents for the university. Most of these folks were big donors to the Bush campaign (a longstanding practice in most states), but I have to say they were prominent and responsible leaders. Governor Bush and his family had great connections throughout the state. Our first experience was that the chairman of the board, a banker from Amarillo, seemed uncomfortable with the research part of the university's mission. The ECDP made appointments and spoke with him, and it had a tremendously positive effect. Being a very bright and conscientious businessman, he quickly learned what we were about. I learned a great deal about the workings of the university by serving on the executive committee. The distinguished professors have played an honorable role in improving Texas A&M University.

Serving on search committees was also an opportunity to learn even more about the university's operations. Upper-level search committees usually included a DP. I served on search committees for provost; deans of Liberal Arts, Agriculture, and Geosciences; department heads for statistics, mathematics, oceanography, and geography; and the CEO of the Galveston Campus. On the statistics search, my friend and collaborator Joe Newton was a candidate. I advised Joe not to do it because it would ruin his research career. After a couple of terms as head, he became a beloved and distinguished Dean of the College of Science for three terms. While Joe was department head of statistics, we cochaired two PhD students, James Harden and Eun Ho Ha, who have gone on to distinguished careers.

Perhaps the most interesting assignment for me came when Ray Bowen was president. President Bowen created VISION 2020 project. Its ambitious aim was to bring Texas A&M up to be among the top twenty public universities in the nation by the year 2020. VISION 2020 had about ten thrusts, such as the one I chaired designated the Graduate Program Theme. We had a rather diverse membership, including a few wealthy donors, one of whom was exercised over whether graduates from Galveston could obtain A&M Rings. I think that guy only attended one meeting.

I shared some of the responsibilities of chair with the dean of graduate studies. His idea of improving the graduate program was to streamline the application process. The point I meant to drive home—and it did prevail—was that the only way to improve our graduate program was to *improve the faculty*. Again, if we were to change Texas A&M for the better, we had to think big—take the responsibility of bringing to Texas the quality comparable to that enjoyed by California universities and a few other states, mostly in the Midwest. The University of Texas was in that class, but Texas A&M lagged far behind. President Bowen was already in agreement with me, and he was able to mold and steer wise recommendations for all the themes in the final report. President Bowen was a great intellectual leader of a campus with so much potential, but a campus unable to perceive it and leap at the opportunity. After retirement, Ray Bowen became a member of the National Science Board, one of the most prestigious assignments that a former university president could occupy. Texas A&M may or may not be among the top twenty today, although some surveys had us there in 2018. There is no question that it has improved itself steadily, and as I look around at the new DPs, I wonder whether I would be elected if I were arriving today.

President Robert Gates contributed mightily to the advancement of Texas A&M. He had more influence in the legislature than the academic intellectual Bowen. Gates brought to the scene his great prestige as former director of the CIA and his ability to deal with politicians of all stripes. He led the university to hire hundreds of new faculty at a time when the job market among public universities was tight. This helped us to hire outstanding new talent. One trivial-seeming but important thing among faculty is parking. A survey was made, and it was found that in the "numbered" parking spots—you pay extra for those—there were often empty spaces. Some bean counter decided that we should gate these areas and sell more permits than there are spots—like the airlines odious overbooking schemes. It was just about to be implemented and this came up during a meeting of the ECDPs with President Gates. It was pointed out to him that people in the "numbered spot" ranks also had many visitors they are carting around. When it rains there are no empty spots: wet visitors. He killed the plan. A keen politician showed his skills among the most political of cultures: academia.

## Department Head and Hiring

During my time as department head (1995–2003), we continued hiring great people to fill the slots left by retirements. There is positive feedback at work: Once you have some great faculty in place, they will have the academic values to maintain quality control on the hiring process. My rule was: *If you want to get better, you have to hire faculty members who are better and smarter than we are.* We have done that steadily, starting with Lee Panetta, Tom Wilheit, Ed Zipser, and John Nielsen-Gammon. It has continued with Ken Bowman, Don Collins, Craig Epifanio, Fuqing Zhang, Renyi Zhang, Ping Yang, Ping Chang, Courtney Schumacher, Andy Dessler, Saravanan, Dick Orville, Mark Lemmon, Istvan Szunyogh, Sarah Brooks, Rob Korty,

Anitia Rapp, Don Conlee, Chris Nowotarski, Tim Logan, Yangyang Xu, and all the rest—not to mention our fine staff providing outstanding support to our mission. After thirty-some years, it leaves me the dumbest guy here.

Hiring good faculty members is always a crapshoot. A young scientist seeking a faculty position in the sciences generally has a PhD from a respectable place and a postdoc of two or more years from an even more respectable place. Respectable here means the candidate has worked with scientists that have international reputations. Their programs are productive and support an environment bursting in creativity. They think big, identifying and chasing important problems to work on. The candidate has been in the company of people who are writing and succeeding with grant proposals. A successful scientist in these situations is one who has special traits. Here is my list: intelligence, creativity in selecting and solving problems (scientific taste), honesty, the desire to be famous and influential, hardworking and sincere, and has social intelligence commensurate with the science community. Most good scientists exhibit most of these traits. It is not necessary to have them all; but if not, one must excel in several to make up for those one lacks.

Often big public universities make poor choices at the hiring and granting-tenure stages. After filtering through these two stages, the success rate is about 50 percent. Even the best universities fail about 25 percent of the time in hiring and granting tenure exclusively to super faculty members. This shows how hard it is to predict outcomes even upon arriving with a fine CV. People evolve, mature, and peak out at different rates. As with professional baseball players, there comes a time in a career when productivity starts to plateau. The passion and desire for prestige finally fade. One becomes satisfied that he/she can climb no farther up the ladder. Turning back is rare. There are exceptions, but it is a myth that a declining professor can be reconstituted as a researcher/scholar. Maybe we have to be satisfied that this is the case, and it is the price we must pay to get very gifted people to endure the long and hard times of graduate school, postdoc and, pretenure days—all adding up to about thirteen or more years of stressful investment. Some are simply burned out once they achieve tenured status. Is it worth it? I say emphatically, "Yes. It is in the public interest to keep these career paths attractive." Even those who peak early can contribute in many ways afterward, but resurrection is usually a myth.

In the Department of Atmospheric Sciences, we have to select candidates that will fit our business model. The model goes like this. We have to have students because the state of Texas pays our salaries and provides the spots for faculty. Our undergraduate program is small (about one hundred majors) because the demand and follow-up jobs are not there—this student load is enough to support about three to five faculty. Hence, we have to have PhD students because the state weights a credit hour at PhD-level in science far more than a nonscience credit hour at lower levels. This means we have to have lots of PhD students to justify ourselves to the administration by generating enough weighted student credit hours. The competition for good graduate students is keen in our field. We have to award the graduate students research assistantships with stipends that are competitive with our peer universities. These assistantships are paid for by government grants—our undergrad offerings do not justify many teaching assistantships. The students, faculty, and other soft money (supported by grants) scientists must publish high-quality work, otherwise, the grant proposals will not be funded. It is all very simple: no or few PhD students = no faculty; no papers = no grants; no grants = no PhD students. Treading this wheel successfully requires a faculty with the good traits enumerated above. Grants from NSF, NASA, NOAA, and DOE carry the most prestige, but funding from anywhere is better than none.

Faculty members in a science department such as ours are very diverse in style, race, gender, specialty, and age (years after PhD). Some work in the field, participating in experiments or data-gathering campaigns in the tropics, polar regions, ships at sea, or chasing severe storms; others are involved in the space program, planning for new missions and uses of data from those spacecraft already in orbit. Some spend their time in front of the computer programming or running model simulations. Some are involved more in teaching and writing. The course of a faculty member's career goes through phases, sometimes visiting several of these roles at different times.

While tending to our business model for keeping the ball rolling, we have to pay attention to balance in the department. Of course, we start with diversity in gender, race, and national origin as reflected in our field and in our state. A diverse group with these features is likely to make better decisions about all kinds of matters. Another diversity issue that may tend to diminish the raising of funds in the model is that of crossing the many subdivisions in atmospheric sciences. For instance, theoretical scientists often do not have a large number of students and therefore do not seek a lot of grant money, but they do contribute mightily to the department's reputation. By contrast, an atmospheric chemist might desire a large number of students to tend to the laboratory work. The business model has to preserve the diverse elements in its distribution.

Early on, the career theme is building a reputation mainly via notable research. In mid-career, it is consolidation; leadership of productive, innovative research groups; and growth of influence. In the final years, many tend to lose some of the passion for fighting and negotiating with referees of papers and proposals. As one ages, it is harder to bounce back from humiliating reviews. At this point, communicating, service, and teaching tend to take precedence—for some, that means attempting to scale the academic, governmental, or corporate administration ladder. Diversity across all axes (academic age, subdiscipline, and other) makes for a better department in fulfilling its mission in the unit's business model as well as in furthering all its missions. It is rare that a faculty member in the latter phases can change to a new specialty or become suddenly more passionate about publishing and funding/guiding graduate students. This lesser productivity in the final years of a career, as indicated by the usual signs, is the price we (our nation) pay to attract people and convince them to go through the stressful years of training and sacrifice.

## Department Achievements

I want to mention a few things I did as department head that I think made a difference in the course of our program. The first was an overhaul of our graduate program. Through a series of (contentious) faculty discussions, we initiated a set of four (later enlarged to six) core courses. These core courses represent the range and depth of the subjects that we think are necessary for a researcher in the atmospheric sciences. We required that students have a B average across these courses. For one thing, we did not want students to obtain a PhD from our department that would reflect badly on us. We wanted the student to have developed a deep understanding of the science and have the ability to conduct research in our field or later in some other. For example, they should know enough and have enough skills to move into areas such as computer science, hydrology, and engineering.

We implemented a comprehensive exam that essentially covers the material in the core courses. We expect our PhD students to master the content of these courses. There is a time limit on when the courses are to be completed and the exam taken—this cannot be dragged out to the point that it is too late to fail them. You cannot keep a student in the dark for too long—it is not fair. Students get a second chance if they fail the exam the first time. The exam includes an oral part. The nature of the oral exam is still a subject of faculty debate—frankly, I am opposed to it. Similarly, exactly how the master's degree fits into the process is continually discussed. A central point of the exam is that it is not graded by the student's graduate committee, but rather by the whole department. Problems on the exam are submitted to the conductor of the exam (most often controlled by the devoted Prof. Lee Panetta). Two separate graders score each problem independently, and the grading and ranking are done in a meeting by the entire faculty.

It is difficult to fail one's own graduate student. This scheme gives our process objectivity and clarity. Students have access to old exams so that they know exactly what to expect. Some prominent schools have dropped such a qualifying exam because of the stress it puts on the students, and its reputation for severity might hurt recruiting—word gets around. But science is often stressful, and it is harmful if a student does not experience competition. Also, we were just beginning to get better. Our students were not so good for the first few years of this experiment. Those prominent departments that dropped their qualifying examination simply did not have the problem faced by a nascent program seeking respectability.

Another change that we made when I became department head was the change of the department's name from *Meteorology* to *Atmospheric Sciences*. As atmospheric chemistry, aerosol particles, radiative transfer, opportunities with the space program, planetary atmospheres, air-sea interactions, and climate science were emerging, we needed to have a title that would tell the community that we were not just interested in what most folks thought "meteorology" meant. The very name *Department of Meteorology* connotes the weather forecaster they see on TV—even my hero President Ray Bowen thought this was our main product—he referred to his admiration for NBC's Al Roker (and of course there was Willard Scott, who got his start as the Ronald MacDonald Clown; my own favorite was

David Letterman). Al, Willard, and Dave were great at their jobs, I do not require degrees in meteorology for TV broadcasters who have learned the basics from experience. But our graduate programs are aimed at future researchers and practitioners, mainly behind the scenes. Indeed, the TV and radio broadcaster presents the public face of our field, but the future of funding and research does not lie predominantly in that narrow sector. Funding agencies need to know the broad range of what we are capable of addressing. There was resistance to my idea, but in the end, it was implemented. I also wanted to make it clear that we embrace faculty with a variety of backgrounds. We have hired faculty with PhDs in chemistry, physics, geophysical fluid dynamics, environmental engineering, meteorology, planetary science, and mathematics.

Two other experiences are worth relating. First is a tenure case where an individual assistant professor was coming through the academic process. Before becoming department head, I was in charge of a committee whose task was evaluating and writing a report on each tenure-track faculty member below the rank of full professor. I would then discuss the finding one-on-one with each faculty member being evaluated. In one particular case, I explained our business model to an assistant professor year after year and documented it in his dossier. He simply could not believe that it was necessary to write scientific papers. He was getting external funding, but I could not convince him that if he did not publish, the funding would end—grant monitors need material evidence, for example, scientific publications, to show their bosses. In the beginning, he was a poor teacher, but improving. This faculty member reached his decision year when I was department head. The tenured faculty voted 8 to 2 to award tenure. I wrote my report to the dean recommending no tenure. I had to take my case to the dean and explain it in detail. After reviewing the case, he agreed with me. The candidate was given another year to find a position. Afterward, not one faculty member came to me in protest of my action. There is a happy ending. This person is very content with a job in a small college where the business model is quite different.

My rule in such matters is "It is OK for the department head (or dean) to turn down a positive vote for tenure or hiring from the eligible faculty, but it is not OK to turn down a negative one," a maxim I picked up way back in St. Louis. Deans and department heads have the ultimate responsibility of quality control at the hiring and promotion decisions.

Collegiality and friendship can interfere with our academic responsibilities. Awarding tenure to a person who will not fulfill his/her role in the business model of the department can lead to disaster over the long term. This should not mean that we ignore such factors as diversity in gender, spousal accommodation, race, sexual preference, or the diversity of expectations of members in different subfields. For example, atmospheric chemists need lots of students to work in their labs, but theorists need few students—in fact, in the latter case, too many students can lead to a decline in the quality of the group. An effective academic administrator must see all of these facets of the individual and of the collective interests of his/her unit. If this is done properly, the unit will maximize its effectiveness in quality and productivity. No one is perfect at forecasting the future behavior patterns of an individual. We simply have to live with this uncertainty, despite our corporate managers above.

As we were in this transition, we found ourselves in a position to hire three new assistant professors. Our tenured faculty numbered below ten at the time. We interviewed several candidates, including a Harvard PhD. Even though we had three openings, we hired no one. I must say the feedback of hiring good people now paid off by their votes in this hiring process. The dean allowed us to keep the open positions even if we did not fill them right away, but he did not know that at the time we were to reject all three candidates. I credit the faculty with the wisdom to pass up the hiring opportunity—I was not so sure of it at the time.

These events early in my time as department head added to those of my predecessor, Ed Zipser, sent out a message to the entire atmospheric sciences community that we were trying to build a great department, and we would not settle for less—we did not broadcast or boast about our deeds, but there are no secrets in the academic grapevine. Indeed, the quality of applicants for our openings improved in the following years. Similarly, the quality of applications for incoming graduate students also improved. Making changes is difficult, and it takes patience. Some members resist for fear of the loss of their own status. The time scale for improving an academic department extends over decades. The ship of academia is slow to turn for the better. But while no one is looking, the same does not hold in the reverse—imagine the rapid slide into mediocrity due to people abandoning ship.

When I came to Texas A&M in 1986, the department had fallen into a rut. But the old-timers were of high quality. They allowed me, Tom Wilheit, Ed Zipser, Lee Panetta, and our mounting energetic newcomers to make these changes with very little resistance. We were extremely lucky in this, and I look back gratefully to our senior friends.

## Brief Self-Study

I have tried to figure out what my goals, personal strengths, and weaknesses actually are. One surely is my skill at selling and especially recruiting. Learning to recruit started in my days at the University of Missouri-St. Louis. We were a new university campus that had no graduate programs. The job market for young PhD physicists was terrible, mainly because of the need for diverting resources to the war in Vietnam, but also because the phase of erecting new branch campuses around the country was ending. Missouri was among the last to wake up to the need for urban and other regional "red brick" campuses. Because other comparable campuses were becoming saturated with faculty, it gave us an edge in hiring. The main attraction we offered was young colleagues such as James, Leventhal, and myself. Also, the old-timers who had founded the department a few years before were not too obstructive. This positive atmosphere and the flexibility to change our innovative curriculum was attractive to candidates. Finally, the campus is located in a middle-class suburb, Normandy, of St. Louis, which happens to be a pleasant city with great restaurants, a zoo, baseball, music, and other amenities. The cost of living is low compared to most big metropolitan areas.

## Consulting Experience

I want to mention a couple of other adventures I had soon after I came to Texas A&M. I mentioned earlier that I had dealt with the contractors at ARC, and I came to know its president, Surendra Anand. Anand started his own company in Maryland, providing well-trained assistants to civil servant-scientists at GSFC. Because of his excellent science background and charm, he was able to hire outstanding young scientists to fill these jobs. Also, the company had excellent benefits, and in general, it was a good place to work, largely without the stress associated with getting papers written and getting proposals written and funded. Many contract scientists did participate at a very high level, as indicated by coauthored papers with their civil servant partners.

When I moved to Texas, Anand suggested that we open a branch office in College Station. I would play the role of consultant. I got approval from my dean to participate (he had a similar role with Shell Development Corporation in Houston). Because of government regulations, I could not do such business with NASA for several more years because of my former affiliation, and certainly not with ARC. Anand opened an office in College Station, and it was named Applied Research Technologies (ART). We soon had enough contracts from the Department of Energy to hire some employees.

We were able to persuade paleoclimatologist and old friend Tom Crowley to join us as the manager of ART. The branch grew and expanded, hiring my former postdocs Kwang-Yul Kim and William Hyde, and oceanographers Matt Howard and Steve Baum. We also hired a masters-level programmer and quality control expert Neil Smith. I had lunch with Crowley several times a week. I stopped by the office afterward sometimes, and my name appeared on some of the proposals and scientific papers. That was my role as a consultant. The arrangement went well since we published a number of good papers over several years. In fact, it was lots of fun.

Tom Crowley and I coauthored the book *Paleoclimatology* during the time he managed the shop in College Station. Oxford University Press published the book as one in a series of books on geology. Our editor, Joyce Berry, told us later that it was the biggest seller in the series. Tom was lead author, and he did most of the work. I was incredibly busy, and he was a better writer than I was. I wrote a few of the chapters, but my role was to make sure the book was accessible to climate scientists from atmospheric sciences and other disciplines relevant to climate science, but with no knowledge of paleoclimatology and its special importance.

After a few years, we closed the ART shop in College Station and let all the local grants and contracts transfer to Texas A&M. Crowley was hired as a research scientist, soon promoted to full professor in oceanography, and the others took soft money positions in either oceanography or atmospheric sciences.

Unrelated to ARC, I also had a short-term contract with the Department of Energy's Pacific Northwest National Laboratory (PNNL) to work on detection of climate signals. It was a tiny consulting contract, but it got me started in this arena and led to a couple of early papers on detecting climate signals. Later we did other papers on that subject. These papers were coauthored with: Kwang-Yul Kim, now full professor at Seoul National University and coauthor of our new book *Energy Balance Climate Models* (North and Kim 2017); Sam Shen, now distinguished professor of mathematics and statistics at San Diego State University; and James Hardin, now full professor and division head of biostatistics in the School of Public Health at the University of South Carolina all joined in.

In the early 1990s, Tom Wilheit and I started a business that would provide algorithms and other software and data products primarily to the aerospace industry. We actually founded a corporation to do this work. It was fun to learn how such a process works. I invested some, but I learned that I would not likely be able to contribute much because of other commitments. I dropped this and all ties to the private sector when I became department head in 1995 to make it clear that I had no outside interests that might compete with the department I was about to lead. Incidentally, I took the job of department head at the age of fifty-seven. I made the crucial calculation that $57 + 8 = 65$. This meant that after two four-year terms as department head, in case my brain had turned to mush, I could at least retire with some dignity. After eight years in this position, I was invited to apply for the dean's job, but I declined. Department head was to be my administrative high-water mark.

## Trends in Academia

I have been a student, a faculty member, and a department head at various universities for over sixty years. I also have had some experience in the private sector and the federal government and as a member of boards of nonprofits. I have observed the university from the point of view of a "distinguished professor" who served many times on university-level committees of all sorts and had frequent contacts with university provosts and presidents. Although many experienced persons have commented on the changes in university operations, I will take yet another crack at it. As discussed earlier, the era of "big science" has had a tremendous impact on universities. The undergraduate experience at big universities has also changed dramatically. College sports were always there, but they have expanded in influence. College spectator sports are a source of leisure entertainment for much-sought-after tuition-paying undergraduates, as well as a lure for funds from well-heeled alums. Local businesses also benefit. Winning at football, baseball, and/or basketball is paramount, which explains the salaries commanded by coaches and their staffs—disclosure: I follow sports, too, since I was an undergraduate. It becomes increasingly tempting to enroll undergraduates that come from the upper-middle and upper classes because of their easily identified glowing credentials and the probability that they will eventually become loyal patrons to the institution. While scholarships and other forms of assistance exist for those less well-heeled, they rarely make up for the cultural and financial problems faced by many applicants.

Earlier I alluded to the business model of a successful science department such as the one I have worked in for over thirty years. Entire universities require a business model as well. It is important that the business model is clearly articulated. For example, if you are a small business owner, you have to go to the bank for loans. The first thing the banker wants to know is your business model—it provides evidence of the likelihood of sustainability.

One imperative is funding. The university as a unit has to support itself. Income can be derived from the following sources: tuition, fees (special charges for classes or disciplines often involving "expensive" labs, etc.), state funding (designated for salaries of teaching faculty, staff related to teaching, etc.), donations, grants and contracts, and local income such as food dispensaries and ticket sales. When I was a student at UTK, the tuition and fees amounted to a total of $225 for three-quarters of classes (with summer off). I worked at a minimum wage job for $0.75 per hour. The same tuition and fees are now $12,624 for in-state residents and the minimum wage is $7.50 per hour—an increase by a factor of ten (unadjusted for inflation), whereas tuition and fees are now fifty-six times higher for in-state students. I am not counting other expenses, because I lived at home for my first year at UTK—estimates of these expenses are available at the present UTK website. There were no "fees" in the 1950s that I can recall. According to the UTK website, an in-state student living on campus with room and board, books, and other expenses, would require $28,562 per year—an out-of-state student would need $46,752.

Average faculty salaries have risen by about a factor of ten (roughly

$12,000 then and $120,000 per annum today). States usually pay a campus according to the weighted student credit hours taught—more weight for science courses and in addition more weight for more advanced courses. When I was department head some fifteen to twenty years ago, the state of Texas paid about $52 per weighted student credit hour. That figure is exactly the same today. The reason for the tuition rise is clear. University administrators easily identified the soft spot: parents of the upper-middle class will pay whatever the flagship public university asks. The state has to rely on votes, and 80 percent of the voters are not in the upper-middle class and above. Sadly, the politicians have been fooling the majority. Raising taxes is a hard sell, but the consequences for the lower 80 percent are not good.

Incidentally, the enormous gap in the price for out-of-state students inhibits the fresh ideas and diversity that students of UTK experience. What a shame. The provincial mindset of students in Tennessee (and Texas, too) could benefit from such exposure. When I was on the VISION 2020 project team, we interviewed business students who were plenty talented and well taught, but they told us it was hard for them to get great jobs because of their limited life experience. I suspect that this situation has improved at Texas A&M because it was called out by the VISION 2020 team. I know of international experiences for undergrads in my own department.

Could a freshman now with a job waiting tables in Knoxville pay for school and make ends meet? No way. The situation is similar in Texas, and in some states, it is even worse. The difference now has to be made up by having much wealthier parents than those in my then lower-middle-class (bottom 40 percent) step up. Many students from lower-income families get a break on their tuition, but still, there are overwhelming living expenses that must be faced. Most middle-class students resort to student loans. Many are not mature enough (in age or culturally) to evaluate the consequences of the debt they rack up. Of course, students who gain a university degree are (on average) destined for higher wages later, but getting through the door and the prospect of repaying their debt is daunting. Upper middles and above don't even have to think about these matters.

Quite a few big public universities simply do not accept these weaker students. They are kicked down the ladder to the less prestigious branch campuses—like the one for which I taught (UMSL) for ten years. A few years ago, I saw statistics that indicated that a large majority of the admissions to Texas A&M came from twenty or so huge high schools in the suburbs of Houston, Dallas, San Antonio, and Austin. There have been great efforts to change this, especially through recruitment of more freshmen from underrepresented groups. A recent report from the James Kent Cooke Foundation (Giancola and Kahlenberg 2016) considers this problem.

Let's return to the business model of a large public university. To keep the ball rolling, the university has to apply a more corporate strategy that makes sure all the flows of funds in and out are balanced (or better). The flows of money from the reneging state legislatures have long ago stalled, so that tuition and fees have had to be increased to right that imbalance, exacerbating the problem just raised. Moreover, research is now a major player in the revenues of the big public universities, and that income becomes multiplied several times as it is spent in local communities. Towns that include big public universities are very attractive and prosperous. Research incomes by some state universities are an important component in that state's economy. Politicians and local leaders are well aware of this income to their regions, just as they are aware of the financial "job-creating" military and government research installations in their districts. Besides the total inflow of cash, the so-called "indirect costs" tacked onto all research grants (sometimes approaching 50 percent at state campuses, higher at private universities) are a factor in the university's budget calculations. These factors are important for the sustenance of the institution—paying for libraries and other facilities demanded by the research population on board. Big public research universities need good researchers who can bring in the bucks—so-called "rain makers." "Publish or perish" is not a joke—not just for professors hoping for tenure, but for the whole institution, town, and congressional district.

It is tempting for state universities to team up with industrial partners as a means of raising revenue. Businesses take advantage of incentives such as labs, new buildings, tax breaks, and other resources that might be dangled. State politicians jump on board with the temptation of "crony capitalism." If a venture fails, the business might walk away and the university gets stuck. I am not so sure anyone benefits from these schemes.

Schools have other constraints besides the ultimate one of cash flow. Membership in prestigious organizations such as the American

Association of Universities is a badge that the rising public university covets—and those that are slipping in status want to maintain. Schools must always be on the lookout for new trends in research so they can poach rising faculty stars from elsewhere to tap those financial and intellectual streams and elevate their prestige indicators.

Faculty membership with tenure is no longer the sinecure it once was. Administrators are finding ways to measure a member's utility through metrics. Administrators seek objective data: the individual's number of papers published (faculty solution: break up your comprehensive papers into a number of "just publishable bits"); the number of citations of his/her papers have garnered (faculty solution: cite every paper you ever wrote and same for your friends; also more papers = more citations); the amount of grant and/or contract income raised (faculty solution: follow the fads of the often lame grant monitors, not your own fancy—no money there); the number of national recognitions through awards (faculty solution: promote yourself and reciprocating friends, endlessly). As I have hyperbolically indicated in the above parentheticals, incentives are ripe for gaming the system to the detriment of science (and the nation's interest)—professors are not that dumb; they figure it out pretty fast. Despite my disparaging the metrics used by the universities, I have to agree that there is some merit to them if used carefully and with balance. In some disciplines (atmospheric sciences is one of them), no two departments are comparable (some include oceanography and/or geology, and maybe geography) in composition and mission. Comparing departments carelessly leads to the "apple and orange" problem.

There are profit-making organizations that contract to survey participating big public universities so that the campuses can be compared objectively. Data inform administrators on how each unit and subunit perform when compared to comparable units and subunits around the country. Information is even parceled out and compared to peers across the country for each individual faculty member in a particular field. Decisions can be made "objectively" based on these comparative data.

"Metrics" are the assignment of numerical scores to the measurable performance of various tasks relative to a mission (see Muller 2009, 2018). Metrics have a place in some contexts, such as completing a large project like the building of a satellite or a ship. In those cases, an array of contractors works in parallel on different aspects of the project. Eventually, a project manager has to merge these threads into the final product. The project manager is a systems engineer who uses metrics to help plan the timing of the merging of various parts. This planning includes early identification of potential bottlenecks, compatibility of components, the inevitable failures of some components, and other factors. Contractors have to be whipped into meeting their milestones and deadlines. Systems engineers are great at this kind of herding. At NASA they are extremely tough individuals, given to loud barking.

I do not want to go on too much about metrics in academia, but in watching the 2017 *Vietnam* documentary by Ken Burns and Lynn Novick, I could not help but notice a parallel with the academic practice of "systems engineering." President Kennedy was keen on building his cabinet with Halberstam's *Best and the Brightest*. JFK selected Robert MacNamara as Defense Secretary. MacNamara came from his civilian job as CEO of Ford Motor Company. He was an expert on systems engineering. In the Vietnam situation, MacNamara employed a great variety of metrics in a kind of systems analysis to measure the progress of the war. He had access to the new computing technology. Hundreds of different metrics were included in his model, the most famous of which were the "body count" and "kill ratio." According to the documentary, it was in the late 1960s that all the data were dumped into the computer model and the timing and outcome of the war were forecasted: The war will be won by the United States in 1965! Unfortunately, it was already 1967. They had tons and tons of numerical data of all stripes. As one of MacNamara's critics in the documentary commented, "When you cannot measure what is important, what you *can* measure becomes important." It reminds me of the cliché often repeated in the science community, "If your only tool is a hammer, you convince yourself that every problem is a nail."

There is something to the idea of the "gut" or gestalt evaluation in arriving at conclusions when you are talking about the future passion that people will have or will not have. The world of humans does not always render itself to "deterministic" solutions, no matter how much data is thrown at the problem. It borders on "historicism," the widely discredited idea that history is determined by initial conditions. Weather is about the same: you can only forecast up to about ten days in advance. (Climate is different—it is not an initial value problem, and unlike economics, it is physics-based). The overdependence on metrics is widespread among all the big universities, both public and private. It may have even begun by infecting the branch campuses imitating the "clever" ideas of the flag-ships.

Personally, I cannot complain of all this—my metrics are OK. After all, this kind of thinking was part of the reason I was brought here, and even how I have lasted so long. This peculiar "survival of the fittest" (and/ or master-gamer) phenomenon has caught on at all the big public univer-sities. Something similar is going on at private institutions, but I know very little about them—my only experience being my two-year postdoc at Penn, and that was in an earlier epoch. It is reminiscent of the times in the early twentieth century when Frederick Winslow Turner started the efficiency movement in the manufacturing sector. This makes for more profit for the routine production line, but might not be best for innova-tion, which is supposed to be the aim of academic science, engineering, and humanities. Japanese manufacturers had better ideas about worker involvement.

Is the new model for academia working out for America? Are money and its friend efficiency playing too much of a role? Are we warping the tenets of Robert K. Merton's sociology of science? Not quite, I think, but there are danger signs. Richard Harris's new book *Rigor Mortis* discusses the horrifying errors coming to light in pharmaceutical research because of the fierce competition and the race for grants and contracts. The funds for the research often come from companies with a market interest in the outcomes of studies, compounding the problems. Richard Harris claims that some research done in the private sector has higher quality in the sense of reproducibility than that conducted in academic settings.

# CHAPTER 12

# GEEZER GIGS

After eight years on the job, I stepped down as department head in 2003 at age sixty-five. I refer to my adventures during the years after I stepped down as "geezer gigs." I was prepared to retire, but I felt pretty good, and my brain seemed to be in fairly good shape—perhaps only a few marbles missing or out of place.

## Endowed Chair

I applied for an endowed chair in the College of Geosciences at Texas A&M with a nominal tenure for five years, renewable once, and it was awarded to me. I argued that I wanted to work on some things that were not likely to be funded by the feds—perhaps I should say "the fads." The Harold Haynes Endowed Chair in the Geosciences was nice, it paid $50,000 per year, as they say, "mad money," for each of the five years. I used most of the money to support my last graduate students and to help others in need in the department. Some was spent on travel to meetings and publication charges. Although allowed, I spent no money on my own salary. The chair was renewable without much fuss after the first five-year term, but I elected to not renew so that a younger faculty member could take it. The fact is, I did not need money; I did not even need the summer salary. Distinguished professors did receive a tenth month of salary for a while during this period but that was eventually terminated—too many DPs and their cut from university endowments was needed elsewhere. Recall even the annual bursary was eventually terminated after many years of service. The entropy of the universe is always increasing, like tuition and parking fees—same, but decreasing with the perks, for professors.

One memorable pleasure of holding the Harold Haynes Endowed

Chair in Geosciences, was that I was able to meet Mr. Haynes, the donor of the endowment funds for my chair, $1 million—one of many such gifts to his alma mater. I visited him twice at his office in the financial district of San Francisco during my attendance at the annual meetings of the AGU in that city. Haynes was probably in his eighties at the time. Our very diligent college development officer Greg Willems (now at Kansas State) made all the arrangements beforehand. I walked from my hotel in the area of the Moscone Convention Center, near Union Square, down Market Street, then off to a skyscraper and up the elevator to the forty-fifth floor. I exited the elevator into a small circular and empty lobby whose circumference was composed of several closed and locked doors. I knocked on the proper one, and a well-dressed receptionist appeared, who delivered me to Mr. Haynes—or "Bill," as he insisted. Bill Haynes was a retired CEO of Chevron and former CEO of Boeing. The receptionist served me coffee, and Bill proved to be a charming conversationalist. He reviewed the framed pictures on his wall for me, which included shots of golfing with George H. W. Bush. He told me that his good friend George Schultz, former Secretary of Treasury and State, had an office in the same suite, but he was currently out of town—too bad, I remain a fan of Schultz's, who has made responsible statements about climate change and what needs to be done about it. Haynes lived around the north side of the Bay, and a driver came to bring him into the office several times a week. After a chat, we retired to a small bar downstairs for a scotch and a light lunch.

I visited him a second time a year or two later. He told me of his many health problems, including multiple surgeries and resections of this or that organ. I shared a few of my own. He was always a generous and courteous gentleman. I was saddened when I learned of his death a few years later.

Now, seventeen years after stepping down as department head, I seem to still be functioning. My research efforts have gradually declined, and I no longer take graduate students, although I am still on graduate committees for several students. In 2014, I stepped down to half time, still teaching my undergrad thermodynamics course and my graduate courses on *Statistics in Climate Science* and *Climate Modeling*. I have taught a five-week summer course (ninety-five minutes per day, five days per week) for about twenty years for graduate students from across the campus. In 2016, I gave up my tenured faculty position and was rehired for two months of salary;

I still teach my summer course, and I schmooze with my colleagues on a regular basis. I come in to the office for about four hours per day. I find it stimulating to spend time in my old office on the twelfth floor with floor-to-ceiling windows facing north with a bird's eye view of the skyline of Bryan. As I hammer away at my computer keyboard, faculty members pop in for lunch, coffee, or just a chat.

A particular joy of my geezer days has been in the service area, sitting on awards committees at the national level, writing books, editing journals, and serving as editor in chief of an encyclopedia. I will use this chapter to describe my participation in some of the assignments I took on in my *afterlife*. I attribute much of my longevity to exercise, eating the right foods, having good doctors, taking the time to read good literature, and most importantly having a home life filled with love and tranquility.

In October 2004, I had a kidney stone for the ages. It was my ninth, and according to X-rays, it was a boulder of diameter five millimeters. You would not want a rock that big to get in your eye or your urethra. Kidney stones go way back for me, as I have mentioned earlier, but this was my biggest. For the first time, this one was too big to pass. My urologist said we would have to crush it by sound—not mere cursing, of course. A truck comes to College Station with all the kidney-stone apparatus about one week per month, and we had to wait a few weeks. Luckily I was not in pain—it had lodged in a comfortable place and was not blocking anything important. Lithotripsy is a means of shattering a kidney stone by sending a shock wave of sound from outside into the body with pinpoint aim. It was first tried successfully in experiments with horses. Anyway, although I was concerned about such a procedure, it only took a few minutes and the thing was gone without pain.

Then in January my same friend and urologist, Dr. Alan Young, informed me that my PSA had leaped—PSA is that indicator of prostate cancer detected in a blood sample, familiar to all old men. He did the biopsy (I will not describe how this is done, but it hurts momentarily). A few days later my cell phone rang as I was driving home from the office. I pulled over and it was Alan, "Three out of ten of your samples taken inside the prostate showed malignancy." The best method at that time was to have surgery to remove the plum-sized problem. It starts with cutting a four-inch vertical incision into my belly. This took some time to heal up given that I was sixty-seven years old. But I did heal up, and in a few months, I resumed exercising and walking. I adopted a new fashion accessory: belts hurt my healing scar, so I went to suspenders. I still wear suspenders today, and I am sure people identifying me say, "Yeah, he's the guy with the suspenders," a sad decline from the former, "Yeah, he's the guy that says, 'Good morning,' in the hallway."

In May of that year, I took a refreshing walk on a Sunday morning of about three miles from our house in Bryan. About two hours after my return, I noticed that there seemed to be the proverbial five-hundred-pound gorilla sitting on my chest, and I was sweating profusely although not hot. I said, "Laura, we have to go to the hospital, I think I am having a heart attack." My diagnosis was correct, as confirmed by the emergency room at St. Joseph's Hospital, about two miles from our home. Dr. Kennon Wigley, a cardiologist, happened to be there at the time, and he scheduled a heart catheterization for the next morning. This meant he would insert a wire (not a coat hanger, but I imagined it to be one) into a blood vessel in my groin and run it up and into one of the secondary arteries around the heart to find where the obstruction was located. Dr. Wigley reamed the artery a bit and inserted a stent. Wide awake, I watched on the monitor as he inserted the stent in one of the secondary arteries around the heart muscle. I was given a card to show in case I ever failed a metal detector test at the airport—it never happened.

After the heart attack, I took drugs for thinning the blood and lowering the blood pressure (one aspirin per day), and lowering the cholesterol—the latter (statin) took several different brands (due to bad side effects) to get the LDLs down. The heart attack had probably caused some damage that led to an arrhythmia a few years ago. That got so bad I had to call on an electrocardiologist in Austin to get a procedure called an "ablation." So far that has worked well for me. I seem to have about fifty other ailments common to old people, but I am getting by.

## Call from Ralph Cicerone

After the heart attack, I continued coming to work, teaching, and serving five years as editor in chief of a journal, *Reviews of Geophysics*, published by the AGU. I was at times forgetful and not quite on my game. Surely it was the pills; eventually, my cardiologist reduced the medications and my head cleared.

The following February (2006), I received a call from Ralph Cicerone, president of the NAS. He remembered me from my service as an assistant editor of the *Journal of Geophysics* when he was its editor in chief, years ago. We also renewed acquaintances when we met in Austin in 2001 to coach the brand-new Secretary of Commerce, Don Evans, on what climate change is all about—he had just learned upon assuming the position that NOAA was in his department.

Ralph explained the situation in his phone call: "We are forming a committee to evaluate the work of Michael Mann and colleagues on the 'Hockey Stick Curve.' We would like for you to serve as chairman of the committee." He went on to explain the circumstances. I told him of the heart attack, but that I felt good and would be happy to serve. He told me that my old friends Mike Wallace and Bob Dickinson, both NAS members, would be on the committee. Mike knew of my health situation, and he would assist me as needed. Later, I learned that Mike was first approached to chair the committee, but he declined because he was so busy, and he had chaired too many such committees recently. He probably suggested me. As we proceeded, Mike was extremely helpful. Bob is a quiet man, but brilliant, insightful, and articulate. Both were indispensable for the success of the committee.

Consider now some history to put the committee's role in perspective. Climate scientists like Phil Jones of the Climate Research Center at the University of East Anglia (UK) and Ray Bradley at the University of Massachusetts were among the first to try using data from so-called "proxies" from the past to construct records of estimates of past surface temperatures averaged over large areas. I have discussed in chapter four and others how such data as the widths of tree rings or seasonal-growth layer thicknesses in tropical coral formations can be used to estimate past environmental conditions. One can count rings or layers demarcating annual growth going back in time to establish the chronology. An indicator such as the width of the ring might serve as a substitute for a thermometer—the width data are referred to as proxies. In some cases, it might indicate some other growth-limiting property such as precipitation in a dry or very wet climate. There are several proxies that can be combined into a blend of indicators. One can calibrate a proxy by comparing it with the instrument record that goes back about a century in many regions. Investigators employ a rather complicated (and delicate) statistical "fitting" procedure to set up the proxy.

Mann, Bradley, and Hughes published a remarkable paper in *Nature* in 1998. The paper featured a graph of the temperature estimates beginning in 1400 CE (CE is current era, the old AD). They published a second paper a year later extending the record back to 1000 CE. The curve resembled a hockey stick with the long handle being the record to about 1800 CE then an abrupt shift upward to the present forming the blade of the hockey stick. The crook in the hockey stick coincided with the beginning of the Industrial Revolution. The graph of the record of atmospheric concentration of $CO_2$ shows almost identical and coinciding shape. The name "hockey stick curve"—attributed to Jerry Mahlmann, then director of GFDL at Princeton—stuck.

A version of the hockey stick curve appeared prominently in the front pages of the 2001 IPCC Report (as shown in the figure nearby). It became a sort of sensation. Climate change advocates waved the graph like a flag. Some took it to mean that the authors had found the "smoking gun" that "proved" the global warming hypothesis. The public would have to learn that one paper does not make a smoking gun, or confirm a paradigm. It is merely one of the early salvos tossed into the scientific meat grinder. Dozens of related papers appeared after the original papers prior to the formation of the Hockey Stick Committee (strictly speaking its formal name was the "Committee on Surface Temperature Reconstructions for the Last 2,000 Years"). These papers examined its methodology, adding new data, condemning it and/or praising it—the usual give and take as in any contact sport. The Hockey Stick Committee had the job of looking into this noise and clarifying the scientific status of the problem for the NAS and the US House of Representatives.

Paleoclimatology rarely reaches the newspapers beyond the science section of the *New York Times*, let alone stirs up a storm like this one. For once, paleoclimatology seemed to have some relevance to public policy. Some were pleasantly surprised at the publicity. The resulting sunshine blew them from their cloistered academic pursuits. Skeptical scientists (not an oxymoron, but actually a redundancy, since all real scientists are skeptical about nearly everything recently appearing in the literature) were more reserved, waiting for the dust to settle. Climate change deniers of all stripes came out of the woodwork with their long knives—occasionally over the top. In retaliation, climate change enthusiasts sprang into action—also, occasionally over the top. Game on! Republicans and

Democrats migrated into bipolar and hardening positions. Even Republicans like Newt Gingrich and John McCain, who had been advocates for climate science, had to scamper home to their bases and become deniers, no doubt contrary to their true beliefs. The iconic hockey stick curve ignited the public's climate wars.

Which side to take was not a tough choice for Rep. Joe Barton (R, TX), chairman of the Commerce Committee in the US House of Representatives. He had once been my own congressman, but after a couple of redistricting actions (some call it "gerrymandering") in Texas, he no longer was. I had seen Barton on a flight once and gave him my card in case he ever wanted to know about climate change from a specialist. He never responded even though he had an office in Bryan/College Station. At the time Barton had the honor of collecting more funds from the oil and gas industry than any other House member. After a few climate deniers told him about the notorious hockey stick curve and its "flawed" science, he stomped into the fray. He wanted serious evidence to bolster what he already "knew" to be the case.

He needed documented information. He ordered Mann, Bradley, and Hughes to turn over all copies of their grant proposals, emails, data, computer codes, and notes to his committee. He extended his hunt to many other climate scientists. He demanded that the NSF—they had funded the research—turn over all of their paperwork on such projects. This stunt, of course, led to a flood of letters to newspaper editors. All three of the principal targets were born brawlers and they fought back. More supporters of climate science, normally on the sidelines, joined in.

Egged on by a team of marginally scientific deniers of climate change, Barton and Rep. Ed Whitfield, head of the Energy Subcommittee, decided that they should assemble a committee to write a report on the matter. They appointed Ed Wegman, a professor of statistics at George Mason University in Fairfax, Virginia, as chairman. Dr. Wegman, who had no experience in climate studies, came up with two members of his committee, one, his former postdoc and another a professor at Rice University in Houston—all PhD statisticians. Compare this with NRC methods of picking a committee for an assessment (to be discussed shortly).

Rep. Sherwood Boehlert (R, NY), chairman of the House Science Committee, admonished Barton (as told to me by Ralph Cicerone), saying, "Joe, that is not the way to conduct such an assessment. We have the

National Academy of Sciences for such tasks." Boehlert referred to the Wegman Committee as "a misguided and illegitimate investigation." At this point, Boehlert approached Ralph Cicerone about conducting such an assessment via a committee set up by the NRC, which is an operational arm of the NAS. Ralph explained all of this in his first phone call to me. He referred to my reputation for objectivity.

Before proceeding with the committee, it is interesting to review some of the deniers and their antics throughout this period. First, let me distinguish between skeptics and deniers. Skeptics have an open mind, and if given sufficient credible evidence, they might accept a new point of view. To me, the term "climate denier" applies to persons who will simply never change their minds no matter what evidence you show them. The border between these two types is fuzzy. Usually, there is a reason that has little to do with science, but may be based upon loyalty to a political party position, a particular interpretation of religious doctrine, and finally profit motives defended by industries that are adversely affected by possible policies derived from the consequences of climate change.

Two Canadians, Steve McIntyre, a retired mining consultant, but with a strong bachelor's degree in mathematics from McGill University, and Ross Mckitrick, an economics professor at Guelf University, undertook to show that the hockey stick was all wrong. Steve's job in the mining industry was to evaluate proposals for new projects. He applied "due diligence" to their requests considering the expenses of mining ventures and the likelihood of success in a venture. I suspect that he was good at his job. Steve started a folksy blog that covered his every thought on the statistics and data quality of the hockey stick papers. He focused most of his attention on tree rings. He made up insulting nicknames for all of his adversaries and managed to gently poke them in the eye at every opportunity. Ross was a professor, but he had some unusual views of science, that made him hard to understand—at least for me. Both guys were always cordial with me when we met face-to-face.

McIntyre and McKitrick even published a paper seeking to debunk the original papers. I had many friendly exchanges via email with them, basically trying to explain climate science—in moderated tones—to them. As with many other email correspondents, I eventually learned what they wanted. It was to read what I had written to them in order to twist it— often out of context—for their own purposes.

They were not alone, a couple of distinguished climate scientists, Peter Huybers of Harvard and Hans von Storch of the Max Planck Institute in Hamburg, separately wrote papers critical of the original papers. By the time the Hockey Stick Committee was composed, more than two dozen papers had appeared in support of the Mann et al. work. These papers used a variety of statistical methods and different choices of proxy data. The scientific process was working in its traditional form. The negative reaction to the original papers stimulated action on all sides. Barton's great stunt had made climate scientists react like cornered wild animals.

Other deniers were busy blowing smoke over the science. Among them were Pat Michaels, at that time the state climatologist for Virginia, and with whom I have always had friendly relations. I even debated Pat at a small Virginia college once. Dr. Fred Singer had a long history of denying scientific findings of public interest, such as the Ozone hole and the connection between smoking and cancer. Fred must have been in his eighties at the time and suggested me to chair committees and speak at denier conferences because he felt that I was at least cordial to climate deniers. Senator Jim Inhofe (R, OK) claimed in the Senate "that man-made global warming is the greatest hoax ever perpetrated on the American people." These advocates were discussed in Oreskes and Conway (2012). More recently, a tweet sent by Donald J. Trump in 2012 in which he claimed, "Global Warming is a hoax propagated by the Chinese," resurfaced during his campaign for the presidency.

## The Committee and Its Work

We were under a very tight temporal constraint. Ralph first contacted me in mid-February 2006. We would have to have a report in print by June 1, 2006, because the House Energy Subcommittee (to the House Commerce Committee) would hold hearings on June 19 and 20. We had to select potential members, vet them, and get them in place in a hurry. We needed a committee of twelve scientists who were diverse in their specialties and not related to any of the principals in the original papers. Members should be able to read and understand the original papers, but not necessarily be working in that same area. I was asked to suggest members and approve some that Ralph or Mike had already suggested. In advance, I mentioned a couple of people I could not allow on the committee, for one or another reason—mainly because of potentially disruptive behavior that might delay our report.

I suggested one climate change skeptic, John Christy, whom I had known for many years. The two of us had even strolled about London one afternoon when we were in Reading for a meeting—I was dragging him off to one of my favorite book stores, Foyle's on the Strand, and he politely obliged. He was nominated and approved for the committee. He took on the chapter in the report on the instrumental part of the surface temperature record. He said that he was skeptical about all the US data sets, and he only trusted the work of Phil Jones of East Anglia, UK. John did a nice job with his chapter. But after the committee's work was done, all the testimony was over with, and our report signed off on, he turned around and would not support its findings.

Deniers wrongfully (and savagely) vilified the same Phil Jones during the "ClimateGate" incident a few years later in 2010. ClimateGate centered on the release of hacked emails that contained some loose language passed in email messages between some climate scientists working on data sets. The release of the hacked messages was timed to coincide with an international meeting of climate scientists and policymakers in Copenhagen. The bogus scandal caused lots of consternation in the public, but like a puff of smoke, it dissipated in a year or so. It was a nasty episode pumped up out of nothing for political purposes.

We had two PhD statisticians on the committee, Doug Nychka of NCAR and Peter Broomfield of North Carolina State University. Both are well respected in their fields, and they had no ax to grind in this affair. One of the most active, informed, and articulate members was Kurt Cuffey, a glaciologist at UC Berkeley. Others included emeritus Yale professor, geochemist, and NAS member, Karl Turekian; Prof. Franco Biondi, a tree-ring expert from University of Nevada, Reno, who wrote the tree ring chapter in the report; Prof. Ellen Druffel, UC Irvine, a specialist in shallow-water ocean cores; Prof. Neil Roberts, University of Southhampton, nominated by me because I was familiar with his book on the last ten thousand years (see Roberts 2014); Dr. Bette Otto-Bliesner, a prominent paleoclimate modeler from NCAR; and of course, Wallace and Dickinson.

We had no time to spare because of the tight schedule—most NRC assessments might be expected to take a year or even two to get their reports out. The mandate from Congress limited our time to complete our task. We had a quick meeting to organize ourselves. The first order of business was to get acquainted and get a feeling for appropriate chapters, lead authors, and so on. We would hold a public hearing at the NAS facilities in DC. The hearing would last two days with invited speakers to present twenty-minute talks on their particular views of the problem of estimating past surface temperatures through statistical inference from proxy data.

We had about twenty speakers at our hearing. Some were superstars, such as Prof. Richard Alley from Penn State, Prof. Daniel Schrag from Harvard, and many others closely tied to the subject of climate reconstructions. The others included the original hockey stick authors and a few skeptics including McIntyre and McKitrick. Tireless denier Fred Singer attended, but he did not have a formal slot (which he protested) on the agenda—this did not stop him from jumping to his feet and commenting often. Michael Mann could not attend on the first day, so his talk was scheduled for 11:00 a.m. on the second day. He showed up just before his talk that lasted over forty-five minutes, even though the session chair tried to limit him to the allotted twenty minutes. As soon as his talk was finished, he looked at his watch and bolted for the door on the grounds that he had to catch a plane—no time for questions. Everyone in the room stared in surprise at each other—so we broke for lunch.

I suppose that Michael Mann was worried that our committee was going to be a firing squad. Had he attended the first day of the hearing and witnessed the overwhelming support by most of the speakers for his work, he might have had a different attitude. However, his strange performance did not affect the committee's deliberations, which stuck to the objective information we had collected.

We had several committee meetings, and the reports back from our reading of the recent literature came in. The hassling and writing began. You would think that it would not be that difficult to bring twelve well-meaning scientists together and come to a consensus. We even argued about what a "consensus" means. Every step before this and after was accelerated because of our deadline. Once our draft was finished, it was sent to thirteen referees, whose names we did not know at the time. The referee reports came back, and we began to work on them—their reports totaled approximately the length of our original document. Two outside scientists were assigned as monitors who would see to it that we addressed every comment and either made a change or gave a reason why we did not. At last we ended with a printed document of 136 pages. There were over four hundred references. We were able to get all of this done in time to submit it for printing and delivery to the congressional committee.

In discussions with Steve Macintyre, I had to explain to him that the committee could not actually do research to repeat the original hockey stick papers—by the way, we all had day jobs. We only had time to try and understand it and to read what other researchers had to say after those first papers were published. I said facetiously that we had to "wing it." From that time onward, Steve referred to me in his blog as "Wing It" North—this, after all the time I had spent trying to help him understand how the greenhouse effect works. What is that old adage about good deeds and punishment? OK, I suppose the guy is sincere.

Before the hearings, our committee members fanned out to meet members of the House and Senate. These assignments were selected by the NAS. They kindly put us in cabs and dropped us at the appropriate entrances to find the member to whom we were assigned. I went to Rep. Barton's office, but he was not in. I do recall that his outer office was decorated with baseball memorabilia. I asked the receptionist if Barton had played for Texas A&M, she said, "No, but he is a big fan." I just remembered this after the horrific shootings of the Republican baseball team in Virginia, just outside DC, in June 2017. Rep. Barton was at the site with his ten-year-old son. Barton was coach of the Republican team. I had a cordial visit with Barton just before our congressional testimony began. We joked, and he assured me that the questions might be tough, but none of it was to be taken as personal.

I also visited Sen. Inhofe's office, where I met with several staffers. One was Mark Morano, now a blogger and climate denier, who frequently sends out humorous emails to a long list of deniers (somehow I get them and occasionally answer his post with a friendly jab of my own—we go back and forth in good humor). Most of the time at that meeting with the staffers at the Senator's office was spent discussing the Washington Nationals and who had recently gotten free tickets. They were fun, but had no interest in doing (or knowing) anything about climate change. I have been to the House and Senate office buildings many times as a UCAR board of trustees member, once even an invitation from Senator Cornyn. In all of these visits, I have never laid eyes on an actual elected individual, only smiley staffers.

## Congressional Testimony

I was introduced and read aloud the statement, which took about fifteen minutes. It can be found online at:
http://www.nationalacademies.org/OCGA/109Session2/testimonies/OCGA_150987
I will include here the summary:

> In summary, as science has made progress over the past few years, we have learned that large-scale surface temperature reconstructions are important tools in our understanding of global climate change. Surface temperature reconstructions are a useful source of information about the variability and sensitivity of the climate system, and they contribute evidence that allows us to say, with a high level of confidence, that global mean surface temperature was higher during the last few decades of the twentieth century than during any comparable period during the preceding four centuries. Further research, especially the collection of additional proxy evidence, would help to reduce uncertainties and allow us to make more definitive conclusions over longer time periods.

My statement opened the testimony, and it was followed by Prof. Wegman's opening statement, then came four hours of questions from the House Committee Members with only one short break in the middle for a light lunch. The questions from (minority) Democrats Stupak and Waxman were well prepared and forceful. Rep. Stupak was originally a policeman who worked his way through law school and then became a prosecutor in Detroit. His aims were partly to (at least appear to) protect automotive jobs in his district, but he was generally supportive in his inquiries. Rep. Waxman was a legendary colorful liberal Democrat, and he was a knowledgeable friend to science. He was severe toward the efforts to thwart scientific work on climate change. Other Democrats on the committee included the Congressperson from Madison, WI, who vigorously supported my position. Finally, my favorite was Democrat Jay Inslee from Seattle. He showed great interest and was particularly interested in ocean acidification, on which he was well versed. After the hearing and after I had returned to Texas, he actually called to thank me for my testimony, and we had a long conversation about climate change and my compliments to the great geoscience programs at the University of Washington. Jay Inslee is now the governor of Washington, not a surprise to me, given his political skills and his intellectual power. He is currently a leading figure in fighting the COVID-19 pandemic.

Republicans on the House Committee were considerably less impressive. One from New Jersey took the opportunity to explain how much research was being done on climate change in his district. A Republican Congressperson from my home state of Tennessee was embarrassing. She was unprepared and even with her twangy drawl, which I suppose only I could interpret, could not have been more grating to my reformed ear. After a few inane comments about climate change, and a few uninformed questions, she came up short for her allotted time and said she would send me a list of written questions to be answered in writing. They did eventually arrive in my mailbox, and they were just as inane and irrelevant as her questions in the hearing. None of the Republicans seemed interested enough in the problem to be very feisty in the hearing. She has just been elected Senator from Tennessee. Wow!

My experience suggests that congressional hearings are showpieces for the committee members to get on the record that they are there for their constituents, donors, and other supporters. At least in my experience, there is very little effort toward the members learning anything—at best, their staffers learned something about climate change. The banter with witnesses is contrived and self-serving. Much of the show is hostile in one way or another, and most of the time at breaks the members assure the witnesses that these are not personal but necessary procedures. It seems phony, but in fact, this is the way democracy works  and in my opinion, the whole thing does work.

A few days after the hearing, the NAS held a press conference at which many well-known reporters appeared. Cicerone introduced me, and I went from there. I looked out at the audience and saw Andrew Revkin of the *New York Times*, Richard Kerr from *Science*, and many other familiar faces. I was tired from the long ordeal and a bit tongue-tied. Fortunately, when I faltered some, up stepped committee member Kurt Cuffey who was beside me at the podium. Also, Peter Bloomfield was eloquent with pertinent remarks on statistics for the reporters. There were favorable reviews in *Science* and *Nature*. Here is a quote from *Nature* (June 29, 2006):

> The academy essentially upholds Mann's findings, although the panel concluded that systematic uncertainties in climate records from before 1600 were not communicated as clearly as they could have been. The NAS also confirmed some problems with the statistics. But the mistakes had a relatively minor impact on the overall finding, says Peter Bloomfield, a statistician at North Carolina State University in Raleigh, who was involved in the latest report. "This study was the first of its kind, and they had to make choices at various stages about how the data were processed," he says, adding that he "would not be embarrassed" to have been involved in the work. Panel members were less sanguine, however, about whether the original work should have loomed so large in the executive summary of the IPCC's 2001 report. "The IPCC used it as a visual prominently in the report," says Kurt Cuffey, a panel member and geographer at the University of California, Berkeley. "I think that sent a very misleading message about how resolved this part of the scientific research was." "No individual paper tells the whole story," agrees North. "It's very dangerous to pull one fresh paper out from the literature."

On the whole, the Hockey Stick Committee was a grand success. After ten years have passed, the climate science community has universally accepted the essence of its findings, even though if you check "hockey stick" on Google, you will find mostly strange references to the Wegman report. Climate deniers are incredibly busy on the internet with gobs of weird interpretations of science. When a climate denier book is published, it immediately rises to the top of the charts because certain organizations buy up copies to jack up the sales enough to get attention on bestseller lists. Other hijinks reminiscent of the dirty tricks in elections are common. If you want to learn more about the controversy, I recommend the

Wikipedia siteshttps://en.wikipedia.org/wiki/Hockey stick controversy and https://en.wikipedia.org/wiki/North_Report, both of which provide historical context and accurate reports of the facts (in my opinion, as of this writing in February 2019). I had no part in writing those Wikipedia articles, and I have no idea who did write them.

In the wake of the hearings, an orchestrated, steady flow of letters to editors and op-ed articles appeared in newspapers and blog posts, vilifying Michael Mann and his colleagues. Ralph Cicerone called me a few times with requests to write letters to editors defending the science community, this was primarily because of "my objectivity as seen in the community." I always responded, and many of my posts did appear in print. I had some experience in this because my colleague Prof. Andrew Dessler and I had coauthored op-ed pieces for publication in Texas newspapers in attempts to educate the public on climate change. Our posts also had drawn many insulting emails to my box. My all-time favorite was "North, you are a whore to the global warming crowd!"

Following is a letter and my response to the *Washington Post*:

## The Right to Question Michael Mann's Climate Research

*Washington Post*, Tuesday, October 12, 2010
Michael E. Mann's Oct. 8 Washington Forum commentary, "Science isn't a political experiment," explained clearly how his ideas on climate science are superior to any accumulation of countervailing facts, and it showed that he's plainly still annoyed that I questioned some of those ideas by holding a public hearing to evaluate them in 2006.

The reality is that the two-day hearing made it clear that Mr. Mann's global warming projections were rooted in fundamental errors of methodology that had been cemented in place as "consensus" by a closed network of friends. The hearing strengthened science because it was informed by various expert work, including that of the National Research Council, which corroborated our central concerns. Mr. Mann's miscalculations would persist today except that they were identified and discussed in public.

Mr. Mann, however, wants to return to the bad old days when nobody was permitted to question the research that drives public policy. He insists that Congress simply do what he says because free debate is troublesome and because anyone who wonders if Mr. Mann got it right can be silenced with derision. I think Mr. Mann is entitled to make up his own mind, but not his own truth.

*Joe Barton, Washington*

The writer, who represents Texas's 6th District in the US House of Representatives, is the ranking Republican on the Energy and Commerce Committee.

Science Committee Responds to Rep. Joe Barton

*Washington Post*, Sunday, October 17, 2010

Regarding Rep. Joe Barton's Oct. 12 letter, "The Right to Question Climate Science," I would like to correct some potential misunderstanding about the conclusions of the 2006 National Research Council report to which Mr. Barton referred. Quoting from the report's summary: "Based on the analyses presented in the original papers by Mann et al. and this newer supporting evidence, the committee finds it plausible that the Northern Hemisphere was warmer during the last few decades of the twentieth century than during any comparable period over the preceding millennium."

While we did find some of the methods used in Michael E. Mann's original papers to be less cautious than some of our members might have used, we have not found any evidence that his results were incorrect or even out of line with other works published since his original papers.

Mr. Barton's reference to "Mr. Mann's global warming projections" is incorrect and quite misleading. Mr. Mann's work does not make projections about global warming. His work, and that of our committee, was concerned with the reconstruction of temperatures in the past. As stated in the report, this area of research does not attempt to make any inference about future temperatures. While knowledge of past climates fills in context, the arguments for anthropogenic global warming are mainly based upon the past fifty years of data, including temperatures, model simulations, and numerous other indicators.

Gerald R. North, Bryan, TX

The writer was chair of the National Research Council's Committee on Reconstruction of Surface Temperatures for the Last 2000 Years, mandated by Congress. The views expressed are his own.

Fake news is not a new phenomenon.

This exchange shows that even four years after the "Hockey Stick" report and the testimony Barton witnessed, the whole kerfuffle had not yet died out. I have to admit that Mann loves the controversy, and he can dish it out as fast as they can deliver it. The attorney general of Virginia tried to subpoena all of Mann's records that were accumulated during his time as a faculty member at the University of Virginia. Much of this occurred after Mann moved to Penn State. I am not sure whether the attorney general ever succeeded in his harassment or not, but Mann got back at him by campaigning on the stump in favor of his opponent and against him when he ran unsuccessfully for governor. I doubt that Mann was a factor in that election, but he showed his mettle and his tenacity.

## Solar Effects Committee

In 2011 I received another call from Ralph Cicerone, this time asking if I would chair the program committee for a project concerning the influence of the solar changes or fluctuations on the Earth's climate. This time NASA HQ asked for a study on this influence because the HQ manager of the Heliosphere Research Program wanted to have a workshop and report on the matter. The heliosphere is the spherical volume of space starting at the Sun's surface and extending out to a boundary that is determined by where the "solar wind" is matched in strength to the galactic winds. It extends to the outer edge of the solar system. Heliospheric science is the study of the fluxes of electromagnetic waves of all wavelengths, particle flows from the Sun that stream out in the solar wind, and the interaction of these phenomena with the atmospheres of the planets.

The program manager needed evidence from climate as well as from interested solar and heliospheric scientists about the effects of solar variability on the climate system and whether she should be funding it. She never revealed this motivation to me, but I always thought it was in her mind as well as that of her superiors at NASA HQ. This was not to be a full-fledged NRC assessment report, but rather a workshop with a short report written by the organizing committee, of which I was to be the chair.

As usual, our first step was to recruit committee members chosen from the climate science, the heliospheric, and the solar science communities. The latter two know each other better than either of them is much acquainted with the climate science community. We were able to recruit Isaac Held from the Geophysical Fluid Dynamics Laboratory (GFDL), and a member of the NAS, to join us on the committee and obtained commitments from a formidable group of participants.

The organizing committee members first met in DC at the NAS building. During a contentious, all-day discussion, we hammered out the agenda for the workshop, and we decided that the location for the workshop would be in the large auditorium on the Mesa site at NCAR in Boulder, Colorado. Clearly different members had different interests in how some subjects were to be emphasized in the workshop and in the final report. Potentially, the report could influence the budget choices made at NASA HQ. In particular, the question was whether impacts on climate science could be used as a justification in funding some helioscience research.

The workshop was a success in terms of finding agreement on the relevant impact in the climate sciences. I learned a lot about the use of paleo information as a clue about ancient solar activity. For example, some Barium isotope abundance in samples is correlated with solar activity. If there is a good chronological indicator in the same record of data, correlations with climate can be estimated to check for a connection. Evidence of this sort is very sketchy and cannot be taken literally on the surface, but it can raise interesting questions. I also learned a lot about the heliosphere, before which I knew little. I knew a fair amount about the Sun because I had read my friend Peter Foukal's book (2013, I had read an earlier edition), in which there were new discoveries that had come to light.

Among the most exciting news was the recalibration of all the past solar radiometers onboard many of NOAA's weather satellites. This had been a problem for decades. Records of estimates of the Total Solar Irradiance from satellites go back to the 1980s. The records showed correlations of the Sun's brightness and the sunspot cycle, but there were "offsets" or differing biases from one satellite to the next. The group at the University of Colorado finally settled the discrepancy by doing a systematic study of the geometry of the individual instruments and making the appropriate adjustments. Now we have a smooth record spanning all the satellites and a new estimate of the Total Solar Irradiance.

The research group at NCAR advanced new ideas about the correlations of sunspot activity to the onset of the El Niño Southern Oscillation cycle (ENSO) cycle. These findings are interesting, but samples were small and some of the climatologists (including yours truly) were skeptical about the conclusions.

After the talks were all over, on the afternoon of September 9, the floor was thrown open for discussion.

The organizing committee stayed around for a few extra days to wrap up the report. As expected, this was a contentious event. I think the main result from the point of view of this memoir on climate science, is that the effects of solar variation on climate are very small and have little or nothing to do with the rapid rise of global temperatures over the last century.

The entire report is available in pdf form at: https://www.nap.edu/catalog/13519/the-effects-of-solar-variability-on-earths-climate-a-workshop

## A Fox News Interview

In roughly 2012, I received a phone call from Fox News, inviting me for an interview. They wanted to do it on the A&M campus. I had no idea what was involved. They explained that they would bring staff from New York, and they had a video contractor in Austin that would bring equipment for the event. I had to clear the desks out of a vacant classroom and be ready for them at 7:00 a.m. on the appointed day. It took them several hours to get set up with all the optics, screens, reflectors, backgrounds, and so forth. They took videos of me walking across the campus with my sport coat, khakis, and straw hat—panama, not cowboy. I must say everyone I talked to was cordial and friendly.

They seated the host—whose name I have forgotten—and me face-to-face under the floodlights; someone straightened my tie and the "tape" started to roll. An image of the hockey stick curve (the one depicted in this chapter) was projected behind and over the shoulder of the host. I think we were under the lights for three hours or so. I answered all the questions and went on with my extended professorial explanations. The host was fascinated. Finally, they thanked me profusely, mentioning what a good job I had done and how much they had learned. They then gathered up all their stuff, again shaking my hand in gratitude. I was waiting (in vain) for hugs. All of my colleagues were impressed with my moment of fame. We informed the extended family to watch for the airing of the show. Maybe I should have my own talk show.

The show was postponed several times and eventually aired at some unusually late hour. The stars of this show were the usual lineup of deniers and fringe skeptics. I remember seeing Bjorn Lomborg, the Danish statistician/economist who seems to be a denier when really he believes in global warming, but thinks it is not important, compared to all the other problems facing us. Then came the parade of deniers, all of whom are familiar to me, McIntyre, and more. I was shown saying a few words and my stroll across the campus with my panama hat. I was probably on for twenty or so seconds altogether in the one-hour show. I was pleased to do this gig, but I am sorry that this cost the network probably a fortune with all the airfare, hotel, and contractor expense. Obviously I filled the cutting room floor knee-deep on that one. Clearly, my position on global warming was not of much interest. So much for my talk show prospects on Fox.

## National Committees

Writing letters of nomination or recommendation are a never-ending but pleasant chore for elder scientists. Requests for letters from former students as well as former and present colleagues just keep coming. I think they will continue until I am ninety. The stream is slowly diminishing, but so are my skills at writing a good letter. Of course, it is flattering to be asked. The same goes for refereeing papers and book manuscripts. The requests are going down, but the time it takes for me to provide an adequate review is going up. I am declining more and more to referee papers.

I received the honor of being named a Fellow of the American Geophysical Union (AGU) in 2004. I was grateful for the recognition of my work over the previous three (now four) decades. A few years later, I was asked to sit on the Union Committee to select the Fellows. My term was for two years. We met as a committee in San Francisco. It is worth a few lines to explain the procedure for becoming an AGU Fellow. The organization has some 62,000 members. The membership is divided into sections, such as atmospheric sciences, planetary science, geology, geophysics, hydrology, and so forth. Nominations for Fellow are submitted to individual sections. For example, the atmospheric science section might be allotted 16 slots to be submitted to the Union Committee. The total for all sections would be 120. The Union Committee would receive the 120 candidates and trim it down to 60, keeping in mind (but not rigidly) the rough apportionments from the original section offerings. Each year the organization inducts 0.10 percent or about 60 members as Fellows.

I also spent three years (the third as chair) on the Research Awards Committee of the American Meteorological Society (AMS). I was on the committee because I had won the Jule Charney Award for research in 2008. Texas A&M has one other AMS national award winner, Courtney Schumacher, who won the Meisinger Award, awarded annually to an outstanding young scientist (under age forty). AMS has a membership of about 13,000. Fellows are awarded to twenty-six members (0.20 percent a year). AMS Fellows in our department include John Nielsen-Gammon, Andrew Dessler, Ping Yang, Renyi Zhang, Ken Bowman, and emeriti Dick Orville, Tom Wilheit, and myself. Ping Yang and Andrew Dessler have won the AGU Ascent Award (for rising young atmospheric scientists).

NASA medals are another bragging point: Tom Wilheit has two; Ping Yang, Mark Lemmon, and I each have one. Ping's and Mark's are special because they were not NASA employees at the time of their selection (mine came while I was at GSFC). Very few outsiders win these medals. A few other awardees included those I was instrumental in hiring, or informally mentoring, such as Fuqing Zhang, whom I hired and who won the AMS Meisinger Award and recently an AGU Fellow, both after leaving us for Penn State. Another was Ed Zipser, whom I essentially hired and who went on to win the Rossby Award and an AGU Fellow after he left us for Utah. I appointed John Nielsen-Gammon to the position of Texas state climatologist, and he did such a good job that he was named a Regent's Professor for his exemplary service.

## Statements on Climate Change

Statements of scientific societies on climate change are another class of geezer gigs. I have participated on committees that hammered out statements from AGU, the American Statistical Association, the American Physical Society, and the American Chemical Society. In the last two, I was conference-called by the more formal committee, since I was not an active member of those societies.

In 2013, I served as chair of the AGU panel to update its statement on climate change. The societies, particularly the geoscience ones, are updated every five years or so. I had been an ordinary member of the one five years earlier—it went smoothly. This one was an ordeal. One climate change denier was on the panel, and he gave us a fit. He kept trying to change the subject from the broad one to focus more on the narrow area that he worked on, to which none of the others would agree. I was the most adamantly opposed, but as chair, I had to work quietly. Nevertheless, in true denier form, he tormented us. Finally, he stopped participating in the conference calls. The panel was very diverse, covering not just pure climate scientists, but a variety of other AGU subdisciplines, such as carbon cycle, hydrology, oceanography, and applied mathematics. One member was a professional writer who had worked in advocacy of climate change policy. Some members objected to this person being on the committee. The phone calls were contentious, everyone had his/her own concept of how the statement might be framed—and detailed. The level of scientific specificity is always a problem in such gatherings. One person insisted that we have footnotes that explain each technical term. It would have been a first in such statements. That contention stayed in the fight to the very last. Finally, I won that one. The mathematician argued for precision on just what we know and do not know. Since I recommended him to be on the committee, I felt it my responsibility to oblige him. In the end, he helped clarify the document. The discussions went on for about eight months before we issued the statement. It can be found at http://sites.agu.org/sciencepolicy/files/2013/07/AGU-Climate-Change-Position-Statement_August-2013.pdf

## Editing

In 2006, I applied for and was selected to be the editor in chief of the journal *Reviews of Geophysics*. The term was for five years. First I had to select editors in the fields of oceanography, space physics, planetary physics, and hydrology. I was to take climate science and atmospheric sciences. The articles in *Reviews of Geophysics* are by invitation only. Each editor and I had to be alert to good contributors in exciting new developments in their field. I had coauthored a review paper long ago explaining all about energy balance climate models (North et al. 1981). I felt it was an honor to be asked, and I think it should be for any geoscientist. The editors have to approach authors, and they often decline because they are so busy. John Dutton, my committee chair when I served on the Board of Atmospheric Sciences and Climate, once said, "If you want something to get done, ask a busy person to do it." When department head, I learned how to get people to do what they would rather not do.

As I was finishing my term with *Reviews of Geophysics*, I was invited to be editor in chief of the *Encyclopedia for the Atmospheric Sciences*, second edition, to be published by Elsevier. The first edition was edited by the late James Holton. I selected one of his editors, John Pyle, Cambridge University, for stratospheric and chemistry articles; I added Fuqing Zhang, Penn State University, for atmospheric dynamics and weather; and I took the other topics: satellite-related studies, turbulence, atmospheric physics, and climate. This project took about four years to complete with more than three hundred articles, a portion of which were unaltered from the earlier edition. This encyclopedia is available in many industrial, government, and research university libraries. The individual articles are usually available to download in pdf form from the library websites.

## Thermodynamics Book

One of the first long-term geezer gigs I undertook sprang from a course I had taught for many years. I was department head at the time, and Bob Runnels, who had taught our undergrad thermodynamics course for many years, suddenly died. I was not planning to teach that semester, but I looked up his notes to see what he had taught. I think his course was a bit more elementary than the one I had in mind. In graduate school, I took a course from Prof. Anderegg that was applicable to both grad and senior undergraduates. Later, when I was a postdoc at Penn, I got to know the author of the book Anderegg used. Prof. Herbert Callen wrote a thermodynamics book that was a new organization of the subject—it clarified all the mysteries that had bothered me on the subject. That book was so original and clear, that I understand it was an essential factor in his becoming a member of NAS.

I imitated Callen's approach to some extent for our atmospheric science majors. I wrote the notes up in LaTeX, a typesetting word processor that made elegant looking equations. I had spiral bound notes printed and available to the students at cost. I taught the course several times, improving the notes as I went. I invited Lee Panetta, Don Collins, and Ping Yang to teach the course using my notes. They gave me advice on improvements. Finally, I asked Tatiana Erukhimova—then a research scientist in our department, now an award-winning teaching-faculty member in physics—to teach the course. She liked the draft and made many comments. This prompted me to ask her to coauthor the notes for publication as a book.

We formed a great partnership in the project. I am a bit too broad-brush as a textbook writer, and Tatiana is careful and attentive to detail, clarity, and style as well as sound pedagogy. She found all of my errors and closed all my gaps. She hauled me into the office on Saturdays and Sundays as we neared the end. Cambridge University Press (CUP) picked up the contract, and the book was published in 2009. I was disappointed with the sales. Frankly, I thought it might have been a bit too difficult for typical undergrads in meteorology—maybe even for the teachers. But CUP assured me that it was a respectable amount of sales for a book of this type. In fact, they have encouraged us to write a second edition. We cannot decide on whether to raise the level and recommend it for graduate students. It needs improvements in the atmospheric chemistry sections, and we are still trying to enlist an atmospheric chemist as the third author.

## TIAS Faculty Advisory Committee

In 2011, Institute Director and Distinguished Professor John Junkins appointed me to the (inaugural) Faculty Advisory Board for the Texas Institute for Advanced Studies. The institute was just ginning up. I served on this board for three years. It was the brainchild of Junkins, with whom I had served on the ECDP many times. John had labored for the formation of this institute for many years. Its purpose was to bring high-level scholars to Texas A&M for limited visits of up to a year or even more. It sought Nobel Laureates and members of the national academies. He had managed to raise a stake of a few million dollars from Chancellor John Sharp to get started. The colleges within Texas A&M were invited to submit nominees for potential visitors.

The vetting process was to be carried out by the board. Distinguished Professor Ed Fry, the associate director from physics, supplied the groundwork on credentials: awards, discoveries, publications, citations, and so forth. The board then used this data in selecting the invitees. I was pleasantly surprised when a large fraction of invitees actually came and were influential in working with our own professors and students. We had several spend visits in our College of Geosciences including Peter Liss, Michael King, and Charles Kolb. There has been a name change for the institute, it is now called The Hagler Institute for Advanced Studies. The reason is that one the all-time great Aggies, Jon Hagler, saw the benefits of such an entity at his old alma mater and made a sizeable contribution to its endowment. I had known Hagler because of his coleadership with President Ray Bowen of the VISION 2020 Project, in which I served as leader of one of the theme committees.

## TAMU Press Faculty Advisory Committee

The director of the Texas A&M University (TAMU) Press appointed me to its Faculty Advisory Committee (FAC) in 2009. I still sit on it. I wondered recently (2017) whether I would be dismissed because I had gone Emeritus, and Director Shannon Davies said, "Of course not, it is a life sentence for you." I cherish those words. The FAC looks at books that the press is considering for publication. These books have already passed the early screening and the refereeing of two or more anonymous outside readers. We see the referee reports (often quite detailed), and the FAC often engages in spirited discussions of every book under consideration. Sometimes they are sent back for modification, but the editors have already done the hard work of bringing the book to nearly final form. It is always a pleasure to attend, participate, and learn in these monthly meetings.

## A Book Club

I have been a member of a (nonfiction) book club called the NAR (NAR comes from seminar, but there is nothing semi- about it) for about seventeen years. The club has been meeting for about forty years—a few of the inaugural members still attend. I am now the second oldest in age of the active members. Meetings occur at a local Chinese restaurant on Fridays at 4:30 p.m. with some social time, and at 5:00 sharp some designated member gives a presentation that runs until 6:00 when everyone ups and walks out. Refreshments include beer and tea, plus a pair of Chinese dumplings halfway through the presentation. Assignments are usually about sixty pages of text. Books are selected by vote, and presenters are volunteers. This membership has brought me in contact with people from many different disciplines and outlooks. We try to avoid discussions of politics and religion. The website indicates books we have read over the years—quite a list. http://people.tamu.edu/~dcarlson/nar/

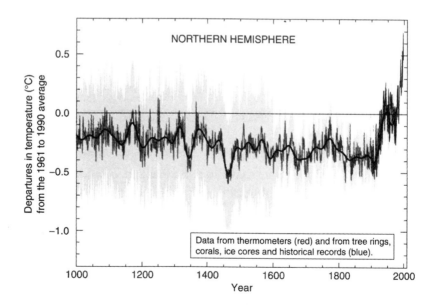

*The image of the Hockey Stick Curve, as shown in the front matter of the Third Assessment Report of the Intergovernmental Panel on Climate Change (IPCC).*

Tom Wilheit and his beloved airplane.

Laura and the boys,
Adam, Paul, and Joel.

262

Laura and Jerry at his eightieth birthday party given by the Department of Atmospheric Sciences.

John Nielsen-Gammon (left) congratulating Jerry (middle) on receipt of the Halbouty Medal by Dean of Geosciences Deborah Thomas (right).

Recent picture of some senior friends in the Department of Atmospheric Sciences: (left to right) Saravanan, Ping Yang, Ken Bowman, Jerry, Andy Dessler, and Renyi Zhang.

North family for Joel's wedding in Washington, DC, May 2018 (left to right) Adam, Paul, Jerry, Michael, and Joel behind Cheryl.

# CHAPTER 13

# THE FUTURE CLIMATE

Climate science has several aims. One is to increase our understanding of this little corner of our world. But there is an engineering side of the science as well. This is the idea of building a framework for projecting our climate system into the future. The task is formidable and not developed to its full potential. But as in medicine and many other fields, we have to do our best at this point with what we have.

## Climate Projections

Through most of this book, I have talked about my involvement in climate science through research, teaching, and service. I also discussed some of the institutions where research in climate science is conducted. I will not claim that I was a major figure in this great project, but I have had a journeyman's role over four decades. There are parts of atmospheric dynamics that still escape me. As with foreign languages, I have a kind of reading knowledge of several, but not a speaking knowledge, especially at the joking and pun level. Now it is time to give readers a short summary of what we know about projecting the state of the Earth's climate into the future and what we can say about the future climate now.

Serious climate forecasts cannot be conducted with the simple but very mathematical models I have spent so many years studying. These models were to give us insight into how the greater system works, but in the end, one cannot forecast the evolution of future climate without a model that computes the detailed circulation of the atmosphere, from global to county-sized spatial scales (a few tens of kilometers), coupled to that of the world oceans across as many scales. Such models have now been in development over the last sixty years. They are still improving, but the improvements are more incremental than those breakthrough leaps in the early days. There are now about twenty of these climate simulation models in operation in laboratories around the world. They are compared to data ranging from observations from countless satellite instruments to those gathered routinely at weather station sites, including ships at sea, for use in weather forecasting.

About every six or seven years, the IPCC issues a report summarizing the state of the art of climate change science. The latest report, Assessment Report Five (AR5), is available online: https://www.ipcc.ch/report/ar5/mindex.shtml

The report is divided into three volumes: Working Group I (WGI) deals with the physical science basis; WGII deals with the impacts of the results of WGI; WGIII deals with the policy alternatives to ameliorate the damages, if any, found in the two previous volumes.

Since I am a physical scientist, and this book is about climate science, I will stick to a summary of what I think is important in the areas covered by WGI. As I emphasized in my congressional testimony to the Energy Subcommittee of the Commerce Committee of the House of Representatives, when it comes to what to do about climate change, "I am not that kind of a doctor—any action taken or not taken will lead to winners and losers in a given region. Deciding on such political choices comes down to moral issues. Moral choices are not scientific facts. I have no more expertise or authority there than a movie star has. I will leave that to our moral leaders, and that includes elected officials, but also others." Of course, I have private views on these moral subjects, but I have no claim to expertise just because I am a climate scientist—very few do. I might have a modest claim because I have read widely on these subjects, but for the most part, I am not a preacher or *political* activist. I am an expert and an activist on the *truth* of climate science.

In preparing a report, IPCC authors meet continuously over the intervals between report deadlines. There is a fair amount of turnover in the composition of scientists participating from one report to the next. The climate modeling teams begin making long computer runs with their latest versions for the next deadline as soon as possible after a report is submitted because of the huge amounts of computer time involved in the preparation of the results for an individual deadline. Frankly, there is not enough time to think and improve things over the short interval available between reports. I have referred in an earlier chapter to the competition between improving the accuracy of simulation versus adding new components such as vegetation and other peripheral effects, even though eventually these have to be accounted for—but not so fast, please!

There are fourteen chapters in the 2013 WGI Report (AR5), each addressing a particular subfield of the project—for example: surface observations, ocean observations, cryosphere (ice), paleoclimatology, and so on. A technical summary document follows these content chapters. Most of this material is written for specialists in the different subfields than for the public. These chapters form an update of the state of the art for participants in the future. For the more general public, there are *Summaries for Policy Makers*, which should provide entry to the more detailed chapters.

In preparing the document, the administrators solicit suggestions for chapter authors. Each chapter is assigned two or three primary authors, and this is supplemented by a large group of coauthors that could number up to thirty. There is an effort to make the coauthors diversified across the 120 or so nations participating as well as across gender, race, and so forth. Workshops are held at the member institutions of the lead authors. I hosted one many years ago at Texas A&M on the problem of "detecting climate signals." That workshop consisted of several days of presentations by the authors and other attending experts on the subject of that particular chapter. Being a primary author is hard work that requires immense travel and unpacking the bickering among the team of writers—I have not done that.

The IPCC makes the draft of an upcoming report open for the public to make criticisms and suggestions once it is finished. An outside oversight team goes over every comment or question received about content. The writing team is required to either implement the suggested changes or provide a reason explaining why not back to the oversight team. A new edition is eventually issued for more public commentary before the final report is sent to press. The early reports were hurriedly prepared and the writing and figures were often poor, but by the time of AR5, the writing was excellent and the graphics were colorful and clear. I have not been an active participant (except for refereeing) for many years, but I know about this evolution because I have taught a climate change class every summer for graduate students for many years and the latest edition of the IPCC report (especially the graphics in the WGI Technical Summaries) is (are) prominently featured in my course.

The method of IPCC differs from those of NAS and the NRC reports. The latter pick a committee typically of about a dozen members. As described in the previous chapter regarding the hockey stick committee, the potential members are carefully vetted for conflicts of interest or for being too close to some of the controversial persons or issues in the assessment that is being given. Then intensive anonymous refereeing is conducted by about the same number of participants. Next, there are two or more monitors who see to it that every objection from referees is addressed or refuted in writing. Independently, the US government also produces periodic reports along the lines of the IPCC reports, called the National Assessments of Climate Change.

There are so many participating countries with diverse political interests in the assembling of the IPCC reports that some concessions have to be made. The authors of the individual chapters are vetted but expertise is paramount; lead authors are sometimes key players in the research on which they are reporting. This practice has provoked criticism for catering to a particular author's pet agenda over strict objectivity. As new reports are issued, the process improves. But lead authors are busy scientists themselves, and the commitment of time, travel, and energy discourages participation in works that return very little credit at home—remember: deans can only count papers and money raised—there is not much dean-worthy recognition for participation in such ventures. More and more, the participants come from government labs rather than academia wherein travel opportunities are more restricted due to teaching schedules. Nevertheless, the IPCC framework is the most massive and thorough scientific assessment process ever undertaken, and in my opinion, it is not perfect, but it probably is about as good as it ever could be, and each release seems better than the previous ones.

After the technical parts of the report are worked out, simplified *Summaries for Policy Makers* are composed. In this last phase, international politics plays an outsized role. Individual countries often attempt to soften or strengthen parts of the content of this section because it might favor the interests of their own country. Sadly, such a process is an inevitable aspect of international diplomacy. Prominent scientist and vocal skeptic Richard

Lindzen has told me that he thinks the science in the WGI reports is excellent; it is the political interference at the end that bothers him. He has a point.

## IPCC Assessment Report Five (AR5) Summary

PROJECTIONS

We cannot make a projection of climate for the rest of the century without taking into account the anthropogenic emissions of greenhouse gases, aerosols, and other changes in conditions that might alter the climate. Hence, the IPCC has developed four Representative Concentration Pathways (RCPs) that climate modelers can use to drive their models in simulations for future climate. These are labeled RCP 2.6, 4.5, 6.0, and 8.5.

Each pathway represents a record into the future of the radiative forcing (mainly greenhouse effect and backscattering of sunlight on aerosols)—a daunting challenge. Recall that the radiative forcing is the imbalance of radiation energy per unit of horizontal area at the top of the atmosphere. For reference, a doubling of $CO_2$ is about 4.0 W/m² (Watts/meter²; the square meter refers to surface area on the Earth's surface). The most "optimistic" of these is RCP 2.6, wherein, the radiative forcing from greenhouse gases and other perturbing influences peaks in about 2040 CE at about 1.8 W/m² from the preindustrial level, and it drops to about half this value over the century. We would love for this case to happen, but it is unlikely.

The RCP 4.5 scenario shows an increase to 4.0 W/m² by 2080 and remains fixed at that value. The RCP 6.0 scenario raises forcing linearly to the end of the century to 6.0W/m², then leveling in around 2150. Finally, the RCP 8.5 scenario ("business as usual") shows 8.4 W/m² by the century's end and levels at 12.4W/m² at 2250 CE. This last scenario is as unthinkable to me as is the first. The two intermediates are probably realistic guides. Unfortunately, human nature sometimes does the unthinkable.

It must be stressed that the uncertainty of climate model simulations is less than the uncertainty of human behavior over the next century as shown in the graphic. An interesting discussion of the difference between model uncertainty and the uncertainty of our own behavior can be found in the *New York Times* article by Justin Gillis:

https://www.nytimes.com/2017/09/18/climate/climate-change-denial.htm
l?rref=collection%2Fsectioncollection%2Fscience&action=click&conte
ntCollection=science&region=stream&module=stream_unit&version=la
test&contentPlacement=3&pgtype=sectionfront

## GLOBAL AVERAGE TEMPERATURE

Once the RCPs (human-action inputs) are given, we can run the models
to estimate the global temperature changes at the end of the century. But
first I must emphasize that the global average temperature change is the
most robust quantity that a climate model can deliver. Because we average over the whole globe, many random errors tend to cancel. But many
details of the heat transport across latitude circles are canceled out in the
global average as well. It is also thought that the global average temperature is most closely connected to the changes in globe-wide average forcings. The reason for this is that the great forcings are also at global scales.

Here are the results for global temperature change at the end of the
twenty-first century: RCP 2.5, the temperature change based on thirty-two
model runs is 1.0 ±0.5°C (1.80 ±0.9°F), above the value at 2000 CE, by the
year 2100 CE. Here the ± sign means the width of the envelope of the individual thirty-two runs (one standard deviation). Using forty-two model
runs, the RCP 4.5 scenario leads to a temperature change of 2.0 ±0.5°C
(3.6 ±0.9°F), above the value at 2000 CE, by the year 2100 CE. Finally,
using thirty-nine model runs, RCP 8.5 yields a change of 4.0 ±1.0°C (7.2
±1.8°F), above the value at 2000 CE, by the year 2100 CE. One thing is
certain: All of the models report that it will warm, some a lot, some less,
depending on the emissions scenario. Differences can also be traced to
the role of the oceans in the scenarios.

## SEA LEVEL RISE

Coming after global average temperature in order of their certainty as
predicted in models (in my opinion) is sea level rise. The reason for my
choosing this indicator next is that the bulk of sea level rise, so far, is due
to the thermal expansion of seawater due to global warming. When the
water is heated, the level will simply rise because the walls (shorelines) are
essentially held fixed. There is a component due to melting ice on glaciers
and that has been rising, but it is still much smaller than the volumetric
expansion. On the horizon, possibly in this century, the melting of ice

sheets such as those on Greenland and Antarctica will become more and more important. The mountain glaciers will be mostly gone by the later part of the century, so their melting effect on sea level rise will diminish.

The ice sheets are more problematic since we do not have sufficient observations and models to make serious predictions, but this is a doable problem, eventually (my opinion). We do have some constraints that help us understand the magnitudes involved. For example, all of Greenland's ice amounts to about six meters of sea level—enough to flood most of Florida. We can look at the map of wastage of ice on Greenland, and we can make an educated guess that at most 10 percent might slide into the sea and melt as icebergs by the end of the century. Our current models project that it will be much less. If we take all of Antarctica's ice, it totals about sixty meters. But most of the ice is in the interior near the South Pole. To the east of the South Pole, the thickness is over four thousand meters, but it is so cold there that no melting is taking place even with the contemplated global warming at the end of the century. The bulk of ice flowing into the sea from Antarctica comes from the West Antarctic Peninsula, the one that points up toward South America. That ice sheet is melting at about the same rate as Greenland.

The RCP 2.6 model simulations project that at the end of the century sea level will rise by 0.45 ±0.15 meters (17.6 ±5.9 inches). The RCP 4.5 and RCP 6.0 results are very similar at 0.47 ±0.15 meters (18.3 ±5.9 inches), while the RCP 8.5 result is 0.625 meters (24.5 ±5.9 inches).

Along many coastlines, especially in the United States, not only is sea level rising, but the coastal land is sinking. Examples include the Chesapeake Bay islands that get into the news occasionally where residents seek aid to building barriers that can prevent erosion of beaches. Land in the neighborhood of Houston, Texas, is sinking partly because fresh water wells have drawn out groundwater with the result of lowering the ground's surface. In other places, the extraction of petroleum has the same effect.

An important issue here is that at least one study I am aware of indicates that even small increases in sea level have an amplified effect on storm surges. Storm surges onshore lead to the most casualties during the landfall of tropical storms. This is likely to be a worsening problem for the Gulf and Atlantic Coastal areas of the United States.

OCEAN ACIDIFICATION

Another robust finding that appears in all model simulations that include ocean chemistry is the acidification of the ocean waters due to the increasing concentrations of $CO_2$ in the atmosphere. We need to do a little chemistry to get this straight. Acidity has to do with the number of free protons that are essentially unbound to their parent molecules in aqueous (salty water is the solvent) solutions. For instance, the compound hydrochloric acid is HCl (hydrogen chloride). When dissolved in water, some of the $H^+$ ion disassociates from the $Cl^-$. These ions move about freely in the water background. A neutral solution is one where the ratio of the number of $H^+$ ions to the background water molecules is $10^{-7}$ (one in ten million), we say the solution has a pH of 7.0. If the ratio is $10^{-6}$ (one in a million, and a pH of 6.0) the solution is a pretty strong *acid*. If the pH is above 7.0 the solution is *basic*.

The average pH of the ocean is slightly basic with a pH of 8.1. It has averaged 8.2 for three hundred million years until the last two centuries, during which it has fallen by 0.1. The concentration of $CO_2$ in the atmosphere equilibrates to its concentration in the oceans. When carbonated soft drinks are bottled, the manufacturer puts an excess pressure of $CO_2$ in the space above the liquid. Some $CO_2$ is dissolved in the drink. If gas pressure is relieved by popping the cap, bubbles will form and $CO_2$ will be released in gaseous form. The solubility of $CO_2$ depends on temperature: hotter water leads to less $CO_2$ in the solution. This is a factor in the storage of $CO_2$ in the cold North Atlantic waters, which is important in the carbon cycle of the Earth system—which we will not get into here. When $CO_2$ dissolves in the ocean waters, it combines with water molecules to form carbonic acid, removing some of the dissolved $CO_2$ and thus "making room" for drawing down more.

Carbonic acid ($H_2CO_3$) poses a problem because it lowers the pH in the ocean waters—one of those $H^+$ ions is free to roam about. Because of the more acidic ocean, some of the ocean's biological balance will be disturbed. The ocean's chemistry will move toward a new equilibrium that is likely to result in erosion of coral and calcium carbonate skeletal materials in the water. We do not know the consequences of this perturbation, which is likely to play out over the next few centuries. What we do know is that life as we know it has evolved over the last three hundred million years under conditions of a slightly basic ocean. Even tiny reductions in pH are likely to upset the ecology of the ocean through evolution and extinction of species. The outcomes will be hard to predict.

Simulations with models that include ocean chemistry yield predictions about the pH ocean waters at the end of the century. The RCP 2.6 scenario leads to a value of 8.05; RCP 4.5 leads to 7.96; RCP 6.0 leads to 7.90; RCP 8.5 leads to 7.75.

PERMAFROST MELTING

Permafrost is water that is frozen year-round below the ground's surface. The soil at the surface of the ground above a permafrost layer may not be frozen in summer, but some distance (a few meters) below the surface, the seasonal cycle is damped out, and it is the mean annual temperature at that site that counts. There is even some permafrost below the sea in the Arctic on the continental shelves. Great areas in Siberia have permafrost. Much of Alaska and Northwest Canada exhibit permafrost. It has been common to see pictures in the news of roads and buildings being tilted and distorted in these areas as permafrost has begun melting. The largest permafrost areas are the Qinghai-Tibet Plateau and the Khangai-Altai Mountains (Central and East Asia). About 24 percent of the land area in the Northern Hemisphere contains permafrost. Hazardous conditions can arise when permafrost on a mountain slope melts because it can result in landslides.

The lower level of permafrost is located where a temperature of 0°C (32°F) is established due to the cooling oozing up the column and the warming due to geothermal heat from below. Thus the permafrost may exist in a layer stretching from a few meters below the surface to depths of more than a kilometer (0.6 mi.).

One concern for climate warming is that as the permafrost melts there can be releases of $CO_2$ and methane because materials locked in the permafrost might now be subject to releases. Typically these deposits have biological origins. Studies are ongoing to assess this grim prospect since it is a (potentially large) positive feedback for global warming.

POLEWARD AMPLIFICATION

Another feature in the global warming scenarios as depicted in climate models is the tendency for warming to increase more in the polar regions than elsewhere. This effect is more prominent in the Northern

Hemisphere than in the Southern. The North Polar region is covered with sea ice. If warming of the Arctic Ocean beneath the sea ice occurs, it will tend to gradually melt the underside of the floating ice, and at some point on the more southerly edges, it can cause the edges of the ice cover to recede poleward. As the sea ice front retreats, the reflectivity of the Earth's surface warms as more solar radiation is absorbed in the dark open water.

Exactly how the water beneath the ice warms is not simple. Water from the outside (Pacific Ocean) gets into the Arctic via the Bering Strait that separates Alaska from Siberia. Stationary weather systems on the Alaska side of the strait set up winds that periodically push water from the Pacific into the Arctic. In any case, the waters in the Arctic do have a mechanism to provide a warm flow under the ice. The sea ice follows wind currents around the Arctic past the Siberian coast with a discharge of icebergs around Greenland and into the Atlantic. There is the possibility of a discontinuous collapse of the summer sea ice altogether, leading to an ice-free Arctic in the summer.

The melting of sea ice in the Arctic has drawn interest in the news because of the likelihood of its eventual summertime demise (perhaps catastrophically; one summer in the future it might go away altogether, but with a thin layer returning in the cold season), allowing for commercial traffic across or around parts of the Arctic Ocean. Another application is the accessibility of the open water for oil and gas exploration. The presence of the ice makes this dangerous and expensive. Obviously, many countries have an interest in the future of this region.

MORE CONTINENTAL HEATING

The climate models uniformly show that the interiors of the great continents will warm a little sooner than the lagging ocean surface waters. This is a result that is found in the simplest climate models, increasing the credibility of the finding. The reason is simple: when an excess of heating at the surface occurs, the land surface lifts its temperature quickly, especially if the soil is dry. If the soil is wet, it takes longer to raise the temperature since some of the heating energy goes into evaporation. Diffusion of heat into the soil is slow and for practical purposes, we can say that the air and the skin of the surface are warmed. Over the ocean, the process is qualitatively different. The upper fifty to one hundred meters of water are well mixed by the winds stirring the surface waters, and this change rapidly spreads down the water column to about fifty to one hundred meters. So when you warm a square meter of the surface, you are warming a column with a thickness of fifty to one hundred meters. This takes time—remember the watched pot that never boils? For this reason, the seasonal temperatures lag behind the heating rates by about three months, while over land it is only a month or so. This helps explain why hurricanes are most frequent in September rather than at summer solstice.

In the case of the slow linear heating of the surface due to the greenhouse effect, the land heats up in temperature much more rapidly than over the oceans. If all forcings were brought to a halt, the temperatures over land and ocean would come to an equilibrium where they are at the same latitude.

## THE MID-LATITUDES AND THE SUBTROPICS

We can think of the mid-latitudes as those from about 30°N to about 50°N (in North America: from roughly the latitude of College Station, Texas, to that of Newfoundland or Quebec; in Europe and Africa: from Algeria or Egypt to the English Channel; and similarly for the Southern Hemisphere).

The zone of the middle latitudes constitutes the "storm belt," that is, where the weather (especially in winter) is punctuated by the passage from west to east of periodic low-pressure areas and fronts bringing precipitation, all following a cadence of about a week's period. This behavior circles the globe, bending some as it goes around because of continentality (large land masses) and topography. Some distortion of the belt also comes from the positioning of shorelines. During autumn, the storm belt marches equatorward, wetting the subtropics just ahead. Then it turns around and marches back poleward during early summer, reexposing the subtropics behind in dryer weather.

Along the equator (±5 degrees of latitude or ±500 km) there is heavy rain, a zone called the Intertropical Convergence Zone (ITCZ). This narrow rainy belt waves back and forth following the Sun overhead through the seasons, especially over landmasses. It is responsible for the annual migration of the grazing beasts in Africa and the great monsoons of India and Southeast Asia. Northern Australia gets a drink of fresh rainwater in its summer.

These giant movements of the atmospheric circulation system wet the subtropics seasonally. In some places, the ocean plays a role, along with the outlines of the continents. But for the most part, the subtropics where so many people live (South Texas, Arizona, New Mexico, Southern California, Mexico, Spain, Italy, Turkey, Israel, and so forth) obtain their water supply from rivers that run from high mountains in the storm belt where giant snow packs store up water and release it in the springtime. These rivers are the very lifesaver for those peoples whose natural habitat is for the most part dry. Consider the Rio Grande, the Colorado, the Nile, the Indus, the Ganges, the Yellow, the Yangtze, the Mekong, just to name a few in the Northern Hemisphere. The snow packs delay the fresh water flowing across the subtropics in these regions, but what if the snow packs are thinner or if they move poleward?

One of the more robust findings of the climate models is that the middle latitude storm belts in each hemisphere are expected to move poleward. This effect is seen in the data as well as in model simulations going back to the 1980s. Simple models for the mechanisms of this shift are not yet satisfactory or well appreciated, but the big models seem to agree. The upshot is that the subtropics should expand and become dryer. This is likely to hit the American Southwest hard, including Northern Mexico. Southern Europe should similarly suffer from this problem.

TROPICS

Our climate model simulations are not so sure about the tropics. The ITCZ might get more intense or might not, depending on the model. Monsoons might get more intense or less. To see how the seasonal cycle operates in the same kind of graphic, see the TRMM homepage: https://pmm.nasa.gov/trmm, or more directly, https://pmm.nasa.gov/trmm/trmm-based-climatology

The African Sahel (that west-east swath just south of the Sahara) may get wetter or not. What about tropical storms and hurricanes (in the Western Pacific, they are called cyclones)? There is meager evidence that they will be less frequent but more intense. The fact is, we have a hard time with simulating tropical convection, which is what all this section is about. The problem boils down to the graininess of the model grid. Convection, especially in the tropics where it can reach up to seventeen km (twelve miles), has a horizontal scale that is still smaller than the grid boxes we can accommodate with present models and computers. We have to fake some of it with (empirically based) fudge factors. The models put the convection in the right places, but the intensity is not in agreement across the population of models. On location (in situ) data are hard to come by; the field experiments are expensive, dangerous, and remote from appropriate facilities. The tropics continue to be a problem in climate science.

## THE CRYOSPHERE

This is the part of the planet where ice exists perpetually—glaciers and ice sheets. The mountain glaciers are melting away rapidly, especially those outside of polar regions (e.g., Kilimanjaro in Tanzania, Quelccaya in Peru, and many others). Many are thought to be finished in forty years. Greenland is melting around its edges as seen from satellite data of several independent types—all in quantitative agreement. Many coastal ice sheets fed by glaciers are extending out over the water on the West Antarctic Peninsula. These ice shelves are breaking away one by one, and as they are set free, they collapse into thousands of icebergs that are dispersed by the circumpolar currents surrounding the continent. When the breakages occur, the glaciers feeding them speed up, leading to more melting and sea level rise.

## EL NIÑO SOUTHERN OSCILLATION AND OTHER OSCILLATIONS

Climate scientists have unearthed many interesting internal oscillations in the ocean/atmosphere system. The foremost is the El Niño Southern Oscillation (ENSO), which I discussed previously. This is a massive change that occurs across the Equatorial Pacific. The cycle begins with a cooling of the waters offshore from Peru and a rise of sea level of a meter or two in the West Pacific. Trade winds across the Pacific push the water westward where it "piles up." Eventually, it fails and a gigantic wave of sea level moves across the Pacific taking months to finally restore the equilibrium. The oscillator is nonlinear, resulting in behavior that is irregular, sometimes recurring in just a few years, sometimes as long as seven years. The warming of the Equatorial Pacific sea surface affects convection along the equator. The dislocation of this convection from the East Indies to the Central Pacific is felt in weather changes all over the world, particularly in North America. What will happen to ENSO as climate changes into the next century? Because of our ignorance of tropical convection and the oceanic subtleties involved, we cannot say. Our current simulations do not show much change.

ENSO is not alone. There are a number of other irregular internal oscillations, such as the North Atlantic Oscillation (NAO), which affects winters in Europe. These are less important global changes than ENSO, but they could be important in some regions of the planet. Climate scientists are still sorting out the various members of the alphabet soup of oscillations (reminds me of the menagerie of elementary particles in my old field of high energy physics) or maybe they should be called fluctuations, indicating that they may be more random than an oscillator might imply. As with ENSO, it is too early to tell what will happen to these creatures (sometimes monsters) in the future.

## UNKNOWN FEEDBACKS

I have alluded to the possibility of hidden feedbacks that we have yet to uncover. The problem is that these are internal adjustments that happen on long time scales—so long that we cannot really simulate them with any confidence. This is in contrast with, say, water vapor feedback that seems to start in just a few days or, as circulation changes occur, in a few months. Actually, even this is not altogether well simulated in the present models.

The long-term feedbacks that we have identified include features of the carbon cycle. Consider this example: $CO_2$ dissolved in sea water can freeze into a solid form called clathrates. A clathrate is a lattice, almost a crystal, of water molecules that can form at low temperatures. Such a lattice can be a "host" and embed $CO_2$ molecules as "guests." These quasi-solid structures sink to the ocean floor and are known to be there in abundance at present. Clathrates can also have methane "guests." Suppose the ocean's waters warm; might these structures come apart and discharge their $CO_2$ or methane into the atmosphere? Some climate scientists have investigated this problem and claim that it would take hundreds or perhaps even thousands of years for such a calamity to unfold.

But how many other feedbacks of this sort do we have yet to find? There are structures like this in the permafrost, and there are ponds in Alaska with methane bubbling out as I write. I am not a doomsday guy, but we should all be aware of such possibilities.

WILD FIRES AND PRECIPITATION

The summer and fall of 2018 saw a huge number of wildfires in the mid-latitudes of the planet. While these were buried in the natural variability of climate fluctuations, that year seems just too extreme in fire phenomena to dismiss. The same is true of extreme rainfall events. The old adage about climate change seems to be holding up: "Where it is dry, gonna get drier; where wet, gonna get wetter." It follows not so much from climate models as from the common-sense idea that when the temperature goes up in dry areas, the likelihood of fires goes up. Similarly, when the air over most area contains more water vapor by thermodynamics, the intensity of rainstorms will increase due to the latent heat release adding to buoyancy. Moreover, the air must be lifted and contain moisture in order for precipitation to occur. The conditions of lifting are found in mid-latitudes where there are mountain ranges along west coasts. In that case, since the west winds across the oceans supply the moisture and the upslopes of the mountains supply the lift, we can expect more rainfall and snowfall. Once the air flows over the mountains, the moisture is removed, and we get the Great Basin of the United States. As the air moves across the country beyond the Rockies, plentiful moisture supplies come from the Gulf of Mexico, and the lift is provided by instabilities in the form of highs and lows. We can expect more drying in the Great Basin and more precipitation in the eastern parts of the United States. These intensifying conditions are on top of the natural variability year to year.

TEXAS CLIMATE AND ITS FUTURE

I teach a course every summer on climate change for graduate students from across the campus. I devote one full ninety-five-minute lecture to Texas climate and its future. Texas's climate is interesting. If we look at the annual averages, we find that the isotherms (curves of equal temperature) run horizontally across the state—cool in the north and warm in the south. On the other hand, the annual average isohyets (equal annual totals of rainfall) run perpendicularly to the isotherms—wet in the east (fifty-five inches in Beaumont) and dry in the west (about seven inches in El Paso). What a stew this could make. Having lived in Bryan/College Station (annual rainfall one meter or thirty-nine inches per year; average temperature 69°F or 20°C; average high temperature in July, August, and September: often over 38°C or 100°F) for over thirty years, I can say that it is cooked to nigh well-done. The temperature from one year to another is quite variable, but normally distributed, and the precipitation variability is much greater, and flatter than a normal distribution. I have seen several years with precipitation less than half the mean. As one moves west to Austin, the annual precipitation falls to a quarter of that at the Gulf Coast and so on to New Mexico.

The North American Monsoon creeps up the west coast of Mexico in late summer to spill a few thunderstorms in the American Southwest—that is when El Paso gets its rain. Otherwise, the western part of the state is (high and) dry.

As we move from the southern tip of the state, Brownsville, to the northernmost panhandle, Amarillo, the elevation goes gradually from sea level to 1099 meters (3605 ft), higher than most peaks in the Appalachian Mountains. This means it cools off faster at night and in summer than in other areas in the state.

Another dominant feature of Texas climate is the wind direction. On average, the wind varies by season, but it comes mostly from the south. This has a profound impact on both temperatures and precipitation. In winter, low pressure areas form as air passes over the Rockies, bringing fronts to Texas. Moist air is brought up from the Gulf of Mexico as the weather features move from west to east across our Great Plains. In West Texas, that air is not so moist, because it comes across the Rio Grande from dry Mexican territories. West Texas heats up; East Texas gets precipitation. Hence, the north/south isohyets line up.

One other feature that is important to Texas climate is ENSO. During El Niño times, the state is a bit warmer and wetter in winter. During La Niña times, the opposite tends to occur. Other patterns of the winter storms crossing the country are affected by sea surface temperature patterns in both the Pacific and the Atlantic. These influential patterns can occasionally conspire to cause extreme climate conditions in Texas that can persist over a few years.

A wonderful novel written by a renowned Texas writer of westerns, Elmer Kelton, called *The Year it Never Rained* describes how ranchers suffered in the record drought of 1957. Pioneers and the generation after them in Texas suffered the vagaries of the intense drought/flood fluctuations as indicated in Robert Caro's volume on the early life of Lyndon Johnson, *The Path to Power*; in Thomas Hatfield's biography of Earl

Rudder, *From Leader to Legend*, telling about Rudder's childhood in the Hill Country; and finally, in the great Texas nature writer John Graves's book, *Hard Scrabble*.

Texas is a pivotal state battered by all the factors listed above with occasional double whammies. Added to these is the gradual warming that is already being felt. Our best guess now is that the American Southwest will become hotter and drier over the course of the century. Precipitation may actually increase in the eastern portions of the state (and nation), but the heating from gradual warming may actually prevail to make the soil dryer. Here we have the differences between large (virtually random) numbers, and this can lead to large variance, which may be larger than at present.

GEO-ENGINEERING

Geo-engineering is a term describing the possible deliberate use of induced climate change to counter global warming. Geo-engineering is being contemplated as a means of adapting to global warming by one of several strategies. Some zealots believe that we can "engineer" our way out of the consequences of increasing greenhouse gases. This would allow us to keep on burning fossil fuels willy-nilly (business as usual) without any of the costs associated with reducing our use of these harmful but convenient substances. Ironically, some of the zealots are folks who told us over the decades that our climate models were incapable of predicting global climate change. Most climate scientists acknowledge our degree of ignorance of future climates, but we feel an obligation to issue the projections along with their uncertainties via the IPCC Reports.

There are several schemes being considered for geo-engineering. One is to capture the $CO_2$ issuing from smokestacks and removing it to some underground storage location. The other is to somehow jiggle the Earth's albedo (reflectivity) to reduce the solar heating of the atmosphere. Presumably, this could be used to offset the increased warming due to our continued use of fossil fuels. The first method (capture and store) is very expensive and has yet to show its feasibility. In fact, there seems at the moment to be a fleeing from coal-fired power plants in favor of natural gas (although a fossil fuel, natural gas is much more environmentally friendly than coal) and the emergence of renewables as an alternative source of energy.

So we should consider the possibility of intervention by changing the albedo of the Earth. There are several schemes being proposed and tested in models, but no serious experimentation has occurred. The two main ways are:

1. Dumping aerosol particles into the stratosphere, mimicking the effects of a volcanic event such as Mt. Pinatubo in 1991. That event sent enough debris into the stratosphere to lower the incoming solar energy flux by $3W/m^2$, three-quarters of the amount doubling $CO_2$ would cause in warming. Rockets could continually spray these otherwise harmless reflecting particles into the stratosphere to cool us down. It would have to be done continuously. Model studies have shown that this would be feasible from an economics point of view. The downsides of this idea include a) it does nothing about the acidification of the oceans, and b) it has to go on indefinitely. This is scary, especially considering the unpredictability of future leaders (even present ones). Could this be used as a military tool? What if a stupid leader sixty years from now were to be a climate denier and claim that the money is being wasted. He/she might stop the spraying into the stratosphere, and now comes the calamity: the globe feels a sudden step up in warming, the havoc of which is well beyond our grasp of climate science today.

2. Making cloud tops more reflective. This could be done, in principle, by making cloud particles smaller and thus more reflective. I find this hard to envision, with ships shooting puffs of salt over the clouds at sea (where the effect would be most appropriate and salt is plentiful). This method also has the drawbacks listed above for the stratospheric intervention.

Ray Pierrehumbert, Oxford professor and climate scientist, raises another objection perhaps even more troublesome, see: http://thebulletin.org/trouble-geoengineers-"hacking-planet" 10858

He is concerned that some well-meaning donors are giving funds to university scientists to conduct in situ experiments on geo-engineering. He feels that this opens a door that cannot be closed. His reasoning is that such donations do not pass the

filter of peer review in the community, and taking this step will
encourage government grant managers (also not subject to peer
review) to open invitations for more programs of this kind. The
danger is that this effort will take funds from other worthwhile
climate science programs and take us down a path that we will
regret because of the *unintended* consequences. I emphasize the
word unintended because it is actually not true: we know about
the consequences and will be tempted to ignore them. The fact
is we do not know enough about climate science to seriously
undertake such reckless programs. I agree with Pierrehumbert's
argument.

## Summary

The Earth will get warmer. How much depends upon the scenario of emissions of greenhouse gases and other anthropogenic factors. The actual outcome should be viewed as one realization out of many possibilities (like the outcome of a single hand-drawn from a deck of cards, but there can only be four aces, etc. Many climate model simulations provide these kinds of constraints.). The models tell us that we can expect a warming over the next century that can be thought of as a bundle of actual outcomes. The mean of the bundle is a few degrees Celsius (multiply by 1.8 to get Fahrenheit). The width of the bundle is about half of that "signal" (we can think of the width or uncertainty as the "noise"). This projected world climate is statistically different from the bundle of outcomes in today's climate. The Earth's temperature is rattling along with a kind of "width of uncertainty." Climate at the end of the century will be more than a "width" of warming.

There are some features of the general circulation of the atmosphere/ocean system that appear to be predictable such as the expansion of the dry subtropical zones of the circulation (called the Hadley Cells). As a consequence, we can say with some confidence that dry subtropical zones will get even dryer, and their poleward edges will move poleward in both hemispheres. Many people live in these arid zones, relying on water from rivers that begin at high elevations in the middle latitudes. This supply of fresh water for people living in the subtropics is likely to be adversely affected by climate change.

Other changes are tied to the change in large spatial scale temperature changes including ice melting and the resulting sea level rise. There are many unknowns about how the oceans will respond. The most robust result is that they will become more acidic, but other uncertainties abound, such as changes in circulation, tropical storms, and sea ice retreats.

For most of society, the outcomes of global climate change are negative, but there will inevitably be winners and losers. The best means of preventing the negative outcomes of increasing $CO_2$ are to decrease the use of fossil fuels, a formidable task. Antidotes have been suggested, but they all seem to attract criticism because of special interests. The recent appearance of inexpensive alternatives to the main culprit, coal, provides some channels of hope through the action of market forces. As ever, I am optimistic.

# CHAPTER 14

# SUMMING UP

This has been a book about a field of science and its growth over the course of my long and somewhat improbable career. Climate science is a field based on the change in paradigm from what I called old-fashioned climatology, perhaps unfairly analogous to stamp collecting, into a physics-based field enabled by the emergence of technologies such as the high-speed and high-capacity digital computer, its friend high-speed information flow, and the ability to observe the planet from artificial orbiting satellites. This has been a phenomenon paralleling other contemporary leaps in science made possible by advances in technology, such as genetics and thousands of others. Such leaps have not only provided for a better-informed world, but they have provided pathways for mobility from all rungs of society into productive and fulfilling lives as working scientists and their associated enabling workforce. The excitement of discovery in science is one of the most exhilarating experiences that one can expect in life. I have had those experiences many times.

Climate science has advanced into a mature phase of normal science where the paradigm of combining global observations range from paleoclimate data retrieval to measurements of Earth's properties from space vehicles, field programs in the most remote parts of the planet to the continuously evolving power of supercomputers and simulations enabled by mathematical advances in numerical techniques. The give and take among observation, experiment, and theory via modeling are alive and constantly advancing our understanding. The public has full access to the findings. Scientists from all nations come together in the endeavor, leading to better science and we hope a better understanding of each other's cultures and interests. We can hope that all of this cross-fertilization will lead to a more peaceful world. We also hope that enlightenment and evidence-based thinking will be enhanced everywhere.

Scientists find it difficult to leave their workbench or computer monitor and share their findings with the public. A huge majority of scientists who work in the field of climate science have agreed for years that most if not all of the projected changes described in the previous chapter are not just plausible, but actually likely. The changes in climate variables (for example, the global warming signal) are now far larger than the uncertainty in model simulations (taken across different models and different runs of the same model—we could call it the signal-to-noise ratio). At this level, one hardly needs a statistician to verify what the naked eye tells us. Anyone can look at the graphs—of sea ice decline, mass-loss of the large and small glaciers, rising sea level, higher temperatures, heavier rainfall in storms in places where it is wet already, more drought where it is already dry, more wildfires, the hockey stick curve, and others—to see that unusual things are happening. There is only one viable explanation—every one of the other possible causes have been tried and rejected.

The solution to the problem is to reduce the use of fossil fuels worldwide. This is well known, but special interests and their influence over less informed people will lead to setbacks. However, I believe that science will triumph in the end. Exactly how do we accomplish this difficult task that will affect some people in the short term as winners and some as losers? I will have to leave that to others—remember I am not that kind of a doctor.

# ACKNOWLEDGMENTS

Many thanks to my family for their unending support. Also, thanks to the countless colleagues and students I've learned from and worked with over my long career. I also wish to thank Shannon Davies, Jay Dew, Emily Seyl, Patricia Clabaugh, and especially Stacy Eisenstark at Texas A&M University Press for their patience and assistance in improving my writing and in making better choices.

# ANNOTATED BIBLIOGRAPHY

## Books and Articles about USSR, Russia

Amster, G., and B. Asbell. 1986. *Transit Point Moscow*. New York: Zebra Books, Kensington Publishing Corp.

Beckwith, C. I. 2009. *Empires of the Silk Road*. Princeton, NJ: Princeton University Press.

Berlin, Isaiah. 1996. *Karl Marx: His Life and Environment*. 4th ed. Oxford: Oxford University Press.

———. 2008. *Russian Thinkers*. New York: Penguin Classics.

Bierly, E. W., and J. A. Mirabito. 1984. "The US–USSR Agreement on Protection of the Environment and Its Relation to the US National Climate Program." *Bulletin of the American Meteorolocial Society* 65: 11–19.

Deutscher, Isaac. 2004. *The Prophet Armed: Trotsky, 1879–1921*. New York: Verso.

Figes, Orlando. 2003. *Natasha's Dance: A Cultural History of Russia*. Reprint. London: Picador.

———. 2014. *Revolutionary Russia, 1891–1991*. New York: Henry Holt & Co.

Frazier, Ian. 2010. *Travels in Siberia*. New York: Picador.

Gaddis, John L. 2012. *George F. Kennan: An American Life*. Reprint. New York: Penguin Books.

Gladkov, F. 1994. *Cement*. Translated by A. S. Arthur and C. Ashleigh. European Classics. Evanston, IL: Northwestern University Press.

Johnson, David. 1998. *The Battle of Britain*. Boston: Da Capo Press.

Kotkin, Stephen. 2015. *Stalin: Paradoxes of Power, 1878–1928*. Reprint. New York: Penguin Books.

Montefiore, Simon Sebag. 2005. *Stalin: The Court of the Red Tsar*. Reprint. New York: Vintage.

———. 2008. *Young Stalin: 1613–1918*. Reprint. New York: Vintage.

———. 2017. *The Romanovs: 1613–1918*. Reprint. New York: Vintage.

Pipes, R. 1974. *Russia under the Old Regime*. New York: Scribner's Sons.

Rasputin, V. 1995. *Farewell to Matyora*. Translated by Antonia W. Bouis. European Classics. Evanston, IL: Northwestern University Press.
Service, Robert. 2011a. *Lenin: A Biography*. Basingstoke: Pan Macmillan.
———. 2011b. *Trotsky: A Biography*. Cambridge, MA: Belknap Press.
Smith, Hedrick. 1976 [1973]. *The Russians*. 1st ed. New York: Quadrangle.
———. 1990. *The New Russians*. New York: Random House.
Thubron, Colin. 2006. *Shadow of the Silk Road*. New York: Harper Perennial.
———. 2007 [1984 in UK]. *Among the Russians*. New York: Atlantic Monthly Press.
———. 1994. *The Lost Heart of Asia*. New York: Harper Perennial.

In addition, all the great novels and stories by Chekov, Dostoevsky, Gogol, Pushkin, Tolstoy, and Turgeniev. My favorite translators are the husband and wife team Richard Pevear and Larissa Volokhonsky. https://www.theparisreview.org/interviews/6385/richard-pevear-and-larissa-volokhonsky-the-art-of-translation-no-4-richard-pevear-and-larissa-volokhonsky

## Memoirs, Biographies, and References about Writing

Ayer, A. J. 2002. *A Life*. New York: Grove Press.
Baker, Russell. 1989. *The Good Times*. New York: William Morrow & Co.
———. 1992. *Growing Up*. Reissue edition. Berkley: University of California Press.
Buenger, W. L., and W. D. Kamphoefner, eds. 2018. *Preserving German Texan Identity: Reminiscences of William A. Trenckmann, 1859–1935*. College Station, TX: Texas A&M University Press.
Caro, Robert. 1982. *The Years of Lyndon Johnson: The Path to Power*. New York: Alfred Knopf.
Catton, Bruce. 2013. *Waiting for the Morning Train*. New York: Doubleday.
Ebert, R. 2011. *Life Itself: A Memoir*. New York: Grand Central Publishing.
Frazier, Ian. 2001. *The Great Plains*. New York: Farrar, Straus and Giroux.
———. 2002. *Family*. Reprint. New York: Farrar, Straus and Giroux.
Gorky, Maxim. 1976. *My Childhood*. New York: Penguin Classics.
Graves, John. 1974. *Hard Scrabble: Observations on a Patch of Land*. Houston: Gulf Publishing Co.

Hamilton, Nigel. 2008. *How To Do Biography, A Primer*. Cambridge, MA: Harvard University Press.

Ignatieff, Michael. 2000. *Isaiah Berlin: A Life*. New York: Vintage.

Josephy, A. M., Jr. 2000. *A Walk toward Oregon: A Memoir*. New York: Alfred Knopf.

King, Stephen. 2010. *On Writing: 10th Anniversary Edition: A Memoir of the Craft Paperback*. Special ed. New York: Scribner's Sons.

Nabokov, Vladimir. 2011. *Speak Memory*. Reissue. New York: Vintage.

Popper, Karl. 2002. *Unended Quest: An Intellectual Autobiography*. 2nd ed. Abingdon, UK: Routledge Classics.

Russell, Bertrand. 1997. *Problems of Philosophy*. New York: Oxford University Press.

———. 2000. *The Autobiography of Bertrand Russell*. 2nd ed. Abingdon, UK: Routledge.

Zinsser, William. 1998. *Inventing the Truth, The Art and Craft of Memoir*. Boston: A Mariner Book, Houghton Mifflin Co.

## Books about Folk History and Opinion

Breeding, Robert L. 1996. *Footprints in Appalachia*. Knoxville, TN: Thriftcon Publications.

Brewer, Alberta, and Carson Brewer. 1975. *Valley So Wild*. Knoxville, TN: East Tennessee Society.

Halberstam, David. 1993. *The Best and The Brightest*. New York: Ballantine Books.

Kephart, Horace. 1913 [and many later printings up to 1995]. *Our Southern Highlanders*. Knoxville: University of Tennessee Press.

McCaslin, R. B. 1997 [1862]. *Tainted Breeze: The Great Hanging at Gainesville, Texas*. Conflicting Worlds: New Dimensions of the American Civil War. Baton Rouge: Louisiana State University Press.

Muller, Jerry. 2007. *The Mind and the Market: Capitalism in Western Thought*. Reprint. New York: Anchor.

———. 2018. *The Tyranny of Metrics*. Princeton, NJ: Princeton University Press.

## Science Biographies and Other Science References

Arrhenius, S. 1896. "On the Influence of Carbonic Acid in the Air upon the Temperature of the Ground." *Philosophical Magazine and Journal of Science* 41: 237–76.

Budyko, M. 1969. "The Effect of Solar Radiation Variations on the Climate of the Earth." *Tellus* 21: 611–19.

Bird, K., and M. J. Sherwin. 2005. *American Prometheus: The Triumph and Tragedy of J. Robert Oppenheimer.* New York: Vintage.

Cassidy, David C. 1991. *Uncertainty: The Life and Science of Werner Heisenberg.* New York: W. H. Freeman & Co.

Chylek P., and J. A. Coakley. 1974. "Analytical Analysis of a Budyko-type Climate Model." *Journal of the Atmospheric Sciences* 32: 675–79.

Colinvaux, Paul A. 1978. *Why Big Fierce Animals Are Rare: An Ecologist's Perspective.* Princeton, NJ: Princeton University Press.

Conrath, B. J., R. A. Hanel, V. G. Kunde, and C. Prabhakara. 1970. "The Infrared Interferometer Experiment on Nimbus 3." *Journal of Geophysical Research* 75: 5831–57.

Crowley, T. L., and G. R. North. 1991. *Paleoclimatology.* Oxford: Oxford University Press.

Donner, L., W. Schubert, and R. C. J. Somerville, eds. 2011. *The Development of Atmospheric General Circulation Models.* Cambridge: Cambridge University Press.

Eckert, Michael. 2013. *Arnold Sommerfeld: Science, Life and Turbulent Times 1868–1951.* New York: Springer.

Eddy, J. A. 1976. "The Maunder Minimum." *Science* 192: 1189–202.

Farmelo, Graham. 2011. *The Strangest Man: The Hidden Life of Paul Dirac, Mystic of the Atom.* New York: Basic Books.

Fisher, D. E. 1988. *A Race on the Edge of Time: Radar; The Decisive Weapon of World War II.* New York: McGraw-Hill Pub.

Fleming, James. 2007. *The Callendar Effect.* Boston: American Meteorological Society.

———. 2016. *Inventing Atmospheric Science: Bjerknes, Rossby, Wexler, and the Foundations of Modern Meteorology.* Cambridge, MA: MIT Press.

Foukal, P., G. North, and T. Wigley. 2004. "A Stellar View on Solar Variations and Climate." *Science* 306: 68–69.

Foukal, P. V. 2013. *Solar Astrophysics.* Weinheim, Germany: Wiley-VCH.

Gauster, W. F., G. G. Kelley, R. J. Mackin Jr., and G. R. North. 1962. "Calculation of Ion Trajectories and Magnetic Fields for the Magnetic

Trapping of High Energy Particles." *Nuclear Fusion Supplement*, Part 1: 239–50.

Ghil, M. 1976. "Climate Stability for a Sellers-type Model." *Journal of the Atmospheric Sciences* 33: 3–20.

Giancola, J., and R. D. Kahlenberg. 2016. *True Merit, Ensuring That Our Brightest Students Have Access to Our Best Universities.* Lansdowne, VA: Jack Kent Cooke Foundation. http://www.Jkcf.org

Greeley, R. S., W. T. Smith, R. W. Stoughton, and M. H. Leitzke. 1960. "Electromotive Force Measurements in Aqueous Solutions at Elevated Temperatures. I. The Standard Potential of the Silver-Silver-Chloride Electrode." *Journal of Physical Chemistry A* 64: 652–57.

Greenspan, Nancy T. 2005. *The End of the Certain World: The Life and Science of Max Born.* New York: Basic Books.

Hallam, Anthony. 2005. *Catastrophes and Lesser Calamities: The Causes of Mass Extinctions.* Oxford: Oxford University Press.

Hays, J. D., J. Imbrie, and N. J. Shackleton. 1976. "Variations in the Earth's Orbit: Pacemaker of the Ice Ages." *Science* 194: 1121–32.

Hegert, W., and T. Wriedt, eds. 2012. *The Mie Theory,* Springer Series in Optical Sciences. Berlin: Springer-Verlag. doi:10.1007/987-3-28783-1_1.

Held, I. M., and M. Suarez. 1974. "Simple Albedo Feedback Models of the Icecaps." *Tellus* 36: 613–29.

Houze, R. A., and A. K. Betts. 1982. "Convection in GATE." *Reviews of Geophysics and Space Physics* 19: 541–76.

James, P. B., and G. R. North. 1982. "The Seasonal $CO_2$ Cycle on Mars: An Application of an Energy Balance Climate Model." *Journal of Geophysical Research* 87: 10271–83.

Kelley, G. G., R. J. Mackin, and G. R. North. 1962. "Calculations of Ion Trajectories and Magnetic Fields for the Magnetic Trapping of High Energy Particles." *Nuclear Fusion Supplement*, Part 1: 80–86.

Kuhn, Thomas. 1996. *The Structure of Scientific Revolutions: 50th Anniversary Edition.* 4th ed. Chicago: University of Chicago Press.

Kullback, S. 1959. *Information Theory and Statistics.* Mineola, NY: Dover Pub.

Leith, C. E. 1975. "Climate Response and Fluctuation-Dissipation." *Journal of Atmospheric Sciences* 32: 2022–26.

———. 1978. "Predictability of Climate." *Nature* 276: 352–56.

Mitchell, J. M., C. W. Stockton, and D. M. Meko. 1979. "Evidence of a 22-year Rhythm of Drought in the Western United States Related to the Hale Solar Cycle since the 17th Century." In *Solar-Terrestrial Influences on Weather and Climate*, edited by B. M. McCormac and T. A. Seliga, 125–43. Berlin: Springer, Dordrecht.

Monk, R. 2013. *Robert Oppenheimer: His Life and Mind (A Life Inside the Center)*. New York: Random House.

North, G. R. 1971. "Solutions of Static Field Problems with Periodic Sources." *American Journal of Physics* 39: 370–72.

———. 1975a. "Analytical Solution to a Simple Climate Model with Diffusive Heat Transport." *Journal of the Atmospheric Sciences* 32: 1301–7.

———. 1975b. "Theory of Energy-Balance Climate Models." *Journal of the Atmospheric Sciences* 32: 2033–43.

North, G. R., R. F. Cahalan, and J. A. Coakley. 1981. "Energy Balance Climate Models." *Reviews of Geophysics and Space Physics* 19: 91–121.

North, G. R., and T. Erukhimova. 2009. *Atmospheric Thermodynamics, Elementary Physics and Chemistry*. Cambridge: Cambridge University Press.

North, G. R., and K.-Y. Kim. 2017. *Energy Balance Climate Models*. Weinheim, Germany: Wiley-VCH.

North, G. R., J. G. Mengel, and D. A. Short. 1983. "A Simple Energy Balance Model Resolving the Seasons and the Continents: Application to the Milankovitch Theory of the Ice Ages." *Journal of Geophysical Research* 88: 6576–86.

North, G. R., F. J. Moeng, T. J. Bell, and R. F. Cahalan. 1982. "Sampling Errors in the Estimation of Empirical Orthogonal Functions." *Monthly Weather Review* 110: 699–706.

Oreskes, N., and E. M. Conway. 2012. *Merchants of Doubt*. London: Bloomsbury Publishing PLC.

Pais, Abraham. 1997. *A Tale of Two Continents: A Physicist's Life in a Turbulent World*. Princeton, NJ: Princeton University Press.

Rhodes, R. 1987. *Making of the Atomic Bomb*. New York: Simon & Schuster.

———. 1996. *Dark Sun: The Making of the Hydrogen Bomb*. New York: Simon & Schuster.

Rudin, Walter. 1996. *As I Remember It. History of Mathematics*, Vol. 12. Providence, RI: American Mathematical Society.

Sellers, W. D. 1969. "A Global Climatic Model Based on the Energy Balance of the Earth-Atmosphere System. *Journal of Applied Meteorology* 8: 392–400.

Stevens, B., and S. Bony. 2013. What Are Climate Models Missing? *Science* 340: 1053–54. doi:10.1126/science.1237554.

Stuiver, M., P. J. Reimer, E. Bard, J. W. Beck, G. S. Burr, K. A. Hughen, B. Kromer, . . . M. Spurk. 1998. "INTCAL98 Radiocarbon Age Calibration, 24,000–0 cal BP." *Radiocarbon* 40: 1041–83.

Weart, Spencer. 2008. *The Discovery of Global Warming*. Rev. and exp. ed. Cambridge, MA: Harvard University Press.

Wilheit, T. T., A. T. C. Chang, M. S. V. Rao, E. B. Rodgers, and J. S. Theon. 1977. "A Satellite Technique for Quantitatively Mapping Rainfall Rates over the Oceans." *Journal of Applied Meteorology* 16: 551–60.

## GRN's Most Cited Papers (Science Citation Index)

Bell, T. L., A. Abdullah, R. L. Martin, and G. R. North. 1990. "Sampling Errors for Satellite-Derived Tropical Rainfall." *Journal of Geophysical Research: Atmospheres* 95: 2195 205.

Crowley, T. L., and G. R. North. 1988. "Abrupt Climate Change and Extinction Events in Earth History." *Science* 240: 996–1002.

Kedem, B., L. S. Chiu, and G. R. North. 1990. "Estimation of Mean Rain-Rate: Application to Satellite-Observations." *Journal of Geophysical Research: Atmospheres* 95: 1965–72.

Kummerow, C., J. Simpson, O. Thiele, W. Barnes, A. T. C. Chang, E. Stocker, R. F. Adler, and K. Nakamura. 2000. "Status of the Tropical Rainfall Measuring Mission after Two Years in Orbit." *Journal of Applied Meteorology* 39: 1965–982.

North, G. R. 1984. "Empirical Orthogonal Functions and Normal-Modes." *Journal of the Atmospheric Sciences* 41: 879–87.

North, G. R., T. L. Bell, R. R. Cahalan, and F. Moeng. 1982. "Sampling Errors in the Estimation of Empirical Orthogonal Functions." *Monthly Weather Review* 110: 699–707.

North, G. R., R. F. Cahalan, and J. A. Coakley. 1981. "Energy Balance Climate Models." *Reviews of Geophysics* 19: 91–121.

Somerville, R. C. J., P. H. Stone, M., Halem, J. E. Hansen, J. S. Hogan, L. M. Druyan, W. J. Quirk, G. Russell, A. A. Lacis, and J. Tenenbaum. 1974. "The GISS Model of the Global Atmosphere. *Journal of the Atmospheric Sciences*31: 84–117.

# INDEX

## Institutions

Other titles in the Kathie and Ed Cox Jr. Books on Conservation Leadership, sponsored by The Meadows Center for Water and the Environment, Texas State University

*On Politics and Parks: People, Places, Politics, Parks*
    George L. Bristol

*Money for the Cause: A Complete Guide to Event Fundraising*
    Rudolph A. Rosen

*Hillingdon Ranch: Four Seasons, Six Generations*
    David K. Langford and Lorie Woodward Cantu

*Green in Gridlock: Common Goals, Common Ground, and Compromise*
    Paul Walden Hansen

*Heads above Water: The Inside Story of the Edwards Aquifer Recovery Implementation Program*
    Robert L. Gulley

*Border Sanctuary: The Conservation Legacy of the Santa Ana Land Grant*
    M. J. Morgan

*Fog at Hillingdon*
    David K. Langford

*The Texas Landscape Project: Nature and People*
    David Todd and Jonathan Ogren

*Wild Lives of Reptiles and Amphibians: A Young Herpetologist's Guide*
    Michael A. Smith

*Discovering Westcave: The Natural and Human History of a Hill Country Nature Preserve*
    Christopher S. Caran and Elaine Davenport